WJTA Conference and Expo 2019

New Orleans, Louisiana, USA
11-13 November 2019

ISBN: 978-1-7138-2818-1

Printed from e-media with permission by:

Curran Associates, Inc.
57 Morehouse Lane
Red Hook, NY 12571

Some format issues inherent in the e-media version may also appear in this print version.

Printed with permission by Curran Associates, Inc. (2021)

For permission requests, please contact WaterJet Technology Association-Industrial & Municipal Cleaning Association at the address below.

WaterJet Technology Association-Industrial & Municipal Cleaning Association
906 Olive Street – Suite 1200
St. Louis, MO 63101-1448
USA

Phone: (314) 241-1445
Fax: (314) 241-1449

www.wjta.org

Additional copies of this publication are available from:

Curran Associates, Inc.
57 Morehouse Lane
Red Hook, NY 12571 USA
Phone: 845-758-0400
Fax: 845-758-2633
Email: curran@proceedings.com
Web: www.proceedings.com

TABLE OF CONTENTS

2019 WJTA Conference and Expo
November 11-13 ● New Orleans, Louisiana

Paper

MODELLING OF THE KERF FORMATION THROUGH PRIMARY AND SECONDARY JET ENERGY FOR THE ABRASIVE WATERJET

E. Uhlmann, C. Männel
Technische Universität Berlin
Berlin, Germany

ABSTRACT

Abrasive waterjet (AWJ) near-net-shape fabrication including AWJ-Milling is a promising non-conventional manufacturing method with advantages for cutting difficult to machine materials. As for AWJ cutting, it is necessary to consider the ratio between the manufacturing time and the workpiece's quality before applying these methods. This ratio is strongly influenced by the AWJ's jet deflection, depending on the AWJ's energy. In this paper a model allowing to predict the kerf formation based on AWJ energy is presented. To study the effects of the AWJ's jet deflection, the AWJ energy is reproduced by a primary and a secondary jet and thus by a primary and secondary material removal. The model was calibrated on titanium aluminide for straight trajectories. The kerf depth and kerf distribution were measured and calculated for curved trajectories. The results reveal that the model reproduces the effects of the kerf formation appropriately. In addition, the model allows to derive strategies to adapt the process and increase the effectiveness of AWJ near-net-shape fabrication.

Organized and Sponsored by WJTA®

1. INTRODUCTION

Abrasive waterjet (AWJ) machining is a non-conventional manufacturing technology, that is mainly used for cutting. The technology inheres a couple of advantages including low heat insertion, low cutting forces, an almost unlimited range of material and no interaction between the tool and the workpiece [1, 2]. The last two points also qualify the technology for machining high-performance materials, which are often considered difficult to machine. Since cutting through the materials limits the attainable geometrical freedom of the technology, AWJ milling has been introduced and investigated during the last years [3, 4, 5, 6, 7]. From the various AWJ milling approaches, maskless AWJ milling allows to create 3D shapes without additional preparations. However, knowing the waterjet's behavior during the manufacturing is of crucial importance to describe the resulting kerf and surface formation. Therefore, models have been introduced to describe the kerf profile depending on the AWJ's parameters [3, 6, 8, 9, 10, 11].

The purpose of models for AWJ milling is to predict the material removal by the AWJ and to foresee the attainable geometrical accuracy. Thus, the expected manufacturing time for a given accuracy can be derived and the costs can be estimated. The estimation of quality and time is particularly important if a decision on a manufacturing process chain needs to be taken. This paper aims to improve the prediction quality and range of materials for AWJ machining. Thus, this model can help AWJ manufacturing to be considered more often when designing new process chains.

1.1 AWJ milling

Most of the AWJ milling models focus on brittle materials and do not consider the effects of secondary material removal by the jet deflection [3, 6, 11]. A model by Van Hung Bui at al. [9] describes the problem of the secondary jet and suggests adapting the sweep pitch in order to minimize the jet deflection (jet escape). Unfortunately, this approach can only be used when cutting even surfaces. Furthermore, a common feature of the models is, that they describe the kerf profile through a bijective line or surface. Thus, the models are not able to predict undercuts, which can appear during AWJ machining and milling. Therefore, it must be assumed, that the models do not describe all effects of the AWJ for all kinds of material in detail.

The fundamental erosion behavior of accelerated particles interacting with a solid material has been described by Finnie [12] and Bitter [13]. They show that the angle of cut α is of crucial importance when the material removal rate is observed. The findings show that for brittle materials the maximum depth of cut appears for an angle of cut of $\alpha = 90°$. Ductile materials on the other side show the highest material removal rate (MRR) around the angle of cut of $\alpha = 20°$ [13].

1.2 Modelling of the AWJ

In this paper, the approach introduced by Axinte [6] is used as basis for the prediction of the kerf profile (jet footprint) and is therefore described in more detail. Axinte [6] presented a geometrical approach of the waterjet based on the assumption that the waterjet's energy and thus the etching rate E correlates with the local waterjet velocity profile $V_{(r)}$ exiting the focus tube, formula 1. Thus, the etching rate E only depends on the process parameters such as water pressure p and abrasive mass flow rate $\dot{m}_A$. The etching rate E is a material specific value. In this paper the etching rate E is referred to as primary etching rate E_l, formula 1.

$$E_l\,(r)= C\,(V_1 \times n_1)^q \qquad \text{Formula 1}$$

The profile of primary etching rate $E_l(r)$ equals the dot product of the waterjet velocity V_l and the normal vector n_l extended with a constant values C and a power factor q which must be calibrated through tests, figure 1. The profile of the primary etching rate $E_l(r)$ can also be derived from the time derivation of a measured kerf profile Z, formula 2. By combining both formulae and transforming the equation to be dimensionless, a nonlinear partial differential equation, formula 3, can be established.

$$\frac{\partial Z}{\partial t} = Z(x,y,t) \cdot n_1 \cdot e_{1z} = \frac{-E_1(r)}{\sqrt{1+\left(\frac{\partial Z}{\partial x}\right)^2+\left(\frac{\partial Z}{\partial y}\right)^2}} \qquad \text{Formula 2}$$

$$\frac{\partial \bar{Z}}{\partial \bar{t}} = \begin{cases} -\dfrac{\varepsilon \bar{E}\left(\sqrt{\bar{x}^2+\bar{t}^2}\right)}{\sqrt[k]{1+\left(\frac{\partial \bar{Z}}{\partial \bar{x}}\right)^2+\left(\frac{\partial \bar{Z}}{\partial \bar{t}}\right)^2}} & \text{for } -\sqrt{1-\bar{z}^2} \leq \bar{t} \leq \sqrt{1-\bar{z}^2} \\ 0 \text{ for } -1 \leq \bar{t} \leq -\sqrt{1-\bar{x}^2} \text{ and } \sqrt{1-\bar{x}^2} \leq \bar{t} \leq 1 & \end{cases} \qquad \text{Formula 3}$$

If high feed speeds v_f are investigated, formula 3 can be further simplified. Afterwards, it is possible to integrate the function to formula 4 which can be inverted to formula 5.

$$\bar{Z}_0(\bar{x}) = -2\varepsilon \int_{\hat{x}}^{1} \frac{\bar{r} * \bar{E}(\bar{r})}{\sqrt{\hat{x}^2+\hat{t}^2}}\, d\bar{r} \qquad \text{Formula 4}$$

$$\bar{E}(\bar{r}) = \frac{1}{\varepsilon\pi}\left[\left[\int_{\bar{r}}^{1} \frac{R(\bar{Z}_0(R)-\bar{Z}_0(\bar{r}))}{(R^2-\bar{r}^2)^{\frac{3}{2}}}\, dR - \frac{\bar{Z}_0(\bar{r})}{\sqrt{1-\bar{r}^2}}\right]\right. \qquad \text{Formula 5}$$

Given a measured kerf profile $Z_0(x)$ it is now possible to calculate the specific etching rate $E_1(r)$. Knowing the specific etching rate $E_1(r)$, formulae 3 can be solved numerically and allows the prediction of the kerf profile $Z_1(x)$ for any feed speeds v_f. Axinte [6] showed that the approach works well for SiC ceramic as target material for feed speeds between v_f = 100 to 1300 mm/min [6].

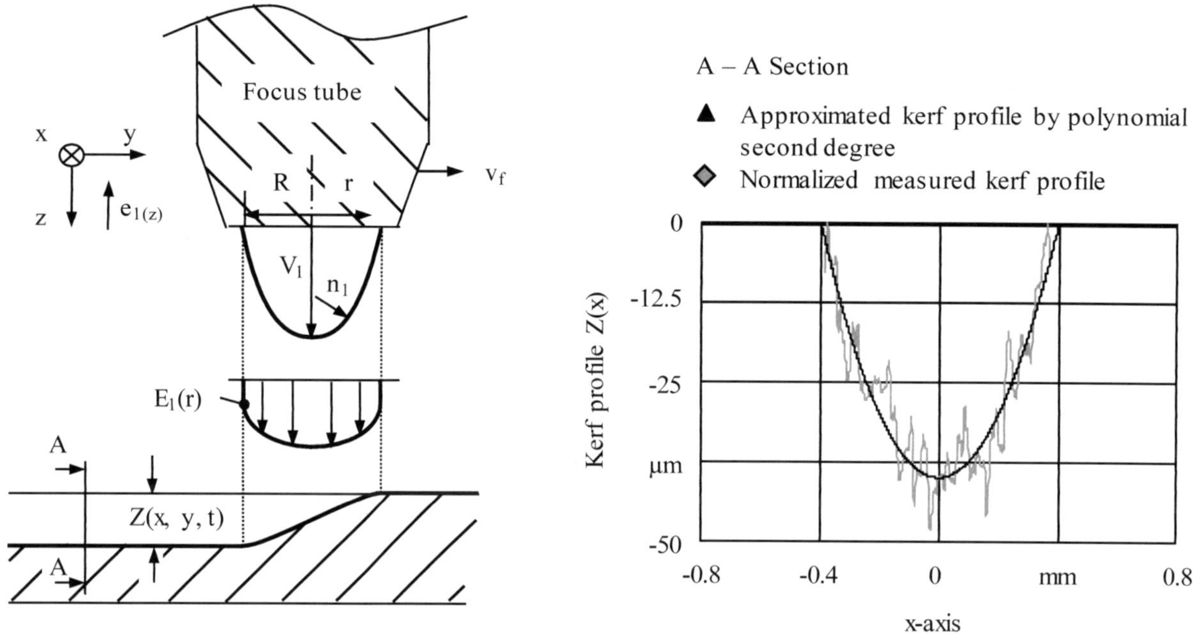

Process:	**Process parameters:**	**Tools:**	**Polynomial values:**
AWJ milling	p = 100.00 MPa	Garnet, Mesh 120, GMA	p1 = 249.37
Workpiece:	$\dot{m}_A$ = 250.00 g/min	d_O = 0.25 mm	p2 = 0.00
TNM-B1 γ-TiAl	v_f = 90.00 m/s	d_F = 0.76 mm	p3 = -39.90
	a_w = 2.00 m	l_F = 76.20 mm	

Figure 1. Primary etching rate $E_1(r)$

2. EXPERIMENTAL AND ANALYTICAL INVESTIGATION

The aim of this paper is to expand the analytical approach by Axinte [6] by reducing its boundary conditions, particularly the affinity to brittle material and the trajectory of the abrasive waterjet. First, the trajectory of the waterjet is generalized. Second, the application to mainly brittle target materials is expanded by implementing a secondary jet energy E_2 leading to a secondary material removal Z_2. This extension aims to describe the effects of ductile materials below v_f = 3600 mm/min. In the end, a further generalization regarding the angle of cut at $\alpha = 90°$ and flat workpiece will be discussed.

In order to calibrate the model several AWJ milling operations have been performed. The tests were carried out on a waterjet machine of MAXIMATOR JET GMBH, Schweinfurt, Germany, type HRX 160 L using a cutting head with a focus tube length l_f = 76.2 mm, a focus diameter

$d_f = 0.76$ mm, an orifice diameter $d_o = 0.25$ mm, a distance between workpiece and focus tube of $l_s = 2$ mm and garnet sand with mesh size 120 of GMA GARNET (EUROPE) GMBH, Hamburg, Germany. The tests and the calculation were implemented for titanium aluminide, type Ti-43,5Al-4Nb-1Mo 0,1B (TNM-B1), of GFE METALLE UND MATERIALIEN GMBH, Nürnberg, Germany.

2.1 Primary Material Removal by the Jet Energy

In order to enable a free movement of the waterjet and variations of the feed speed v_f during a cut, the calculation of the kerfs profile was implemented using MATLAB Release 2015b, the MATHWORKS, INC., Natick, United States. A stepwise algorithm was implemented which assigns a dwell time t_d to every position of the waterjet P_{WJi} along a trajectory. The dwell time depends on the feed speed v_f and the control time t_c. Variations of the control time t_c allow an adjustment towards more precise predictions, or faster calculations. The waterjet positions were extracted from a simple G-Code list allowing the definition of waterjet positions P_{WJ}, feed speeds v_f, radii of the waterjet's trajectory R_{WJ} and the corresponding radii direction (G02, G03), figure 2. Once the waterjet movement and the dwell times t_d were calculated, the effect of the etching rate $E_1(r)$ towards a surface was established. The applied surface consisted of x-, and y-positions between an upper limit UL and a lower limit LL and a defined distance between each point, resolution $res_{i,j}$. All points were defined with a starting z-value of $Z_{1ij} = 0$. The effect of the jet energy was calculated by checking the distance between all points on the surface to the position of the waterjet P_{WJi} for every time step t. If the distance was smaller than the maximum radius of the waterjet $r < R$ the z-value was reduced according to the jets local etching rate $E_1(r)$, formula 6.

$$-\varepsilon_{allg}\bar{E}_1(\bar{r}) = -\frac{v_{f1}}{v_{f2}} \cdot \frac{1}{\pi} \cdot \frac{(p1 - p3 - 2 \cdot p1 \cdot r^2)}{\sqrt{1 - r^2}} \qquad \text{Formula 6}$$

The local etching rate $E_1(r)$ has been found according to formula 6. This formula was derived from a measured and simplified kerf profile $Z_0(x)$, for a feed speed of $v_f = 5400$ mm/min, pressure $p = 100$ MPa and an and abrasive mass flow rate of $\dot{m}_A = 250$ g/min. In this paper the kerf profile $Z_0(x)$, formula 7, of a straight trajectory was assumed to be a polynomial second order. Thus, the kerf profile $Z_0(x)$ could be defined by two parameters p1 and p3. If the profile is symmetric around the x-axis and dimensionless p1 = -p3. This approximation allowed a good description of the kerf with a coefficient of determination $R^2 = 0.98$, figure 1.

$$Z_{0(x)} = p1 \cdot x^2 + p3 \ (-1 \leq x \leq 1) \qquad \text{Formula 7}$$

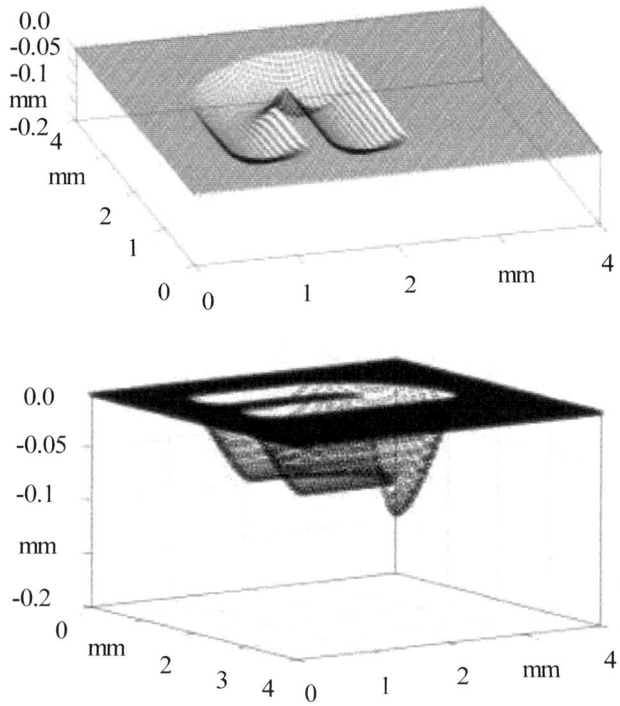

Process:
AWJ milling
Tools:
Garnet, Mesh 120, GMA
d_O = 0.25 mm
d_F = 0.76 mm
l_F = 76.20 mm
a_w = 2.00 mm
Workpiece:
TNM-B1 γ-TiAl
Process parameters:
p = 100.00 MPa
$\dot{m}_A$ = 250.00 g/min
v_f = 30.00 m/min

Calculation parameters:
t_c = 0.08 ms
R = 0.40 mm
UL_x = 4.00 mm
LL_x = 0.00 mm
res_x = 0.05 mm
UL_y = 4.00 mm
LL_y = 0.00 mm
res_y = 0.05 mm
p1 = 39.90 µm
p3 = -39.90 µm

G-Code input file

Pos.:	X	Y	v_f	R_{WJ}	G
0	1	1			
1	1	2	30		
2	2	3		1	2
3	2	1			

Figure 2. Results of the primary etching rate $E_1(r)$ for a free trajectory

Applying the presented approach, the kerf profile can be calculated for any feed speed v_f and waterjet trajectory above a surface. Figure 2 shows the movement of the waterjet along the trajectory given in the G-Code file in the figure. A straight path, a radius and a sharp corner were implemented. It is possible to observe a deeper maximum depth of cut d_c close to the sharp corner. This observation provided prove for the functionality of the approach, since this phenomenon is known, described and discussed by Laurinat [14].

2.2 Secondary Material Removal by the Jet Energy

The approach using a primary jet energy orthogonal to the surface allows a good prediction of the kerf formation, especially for brittle materials and high feed speed v_f [6]. Figure 3 shows that the approach works as well for high feed speeds on the ductile material TiAl TNM-B1. However, below a feed speed of $v_f = 3600$ mm/min a difference between the calculated $Z_1(x = 0)$ and the measured $Z_0(x = 0)$ kerf profile can be observed.

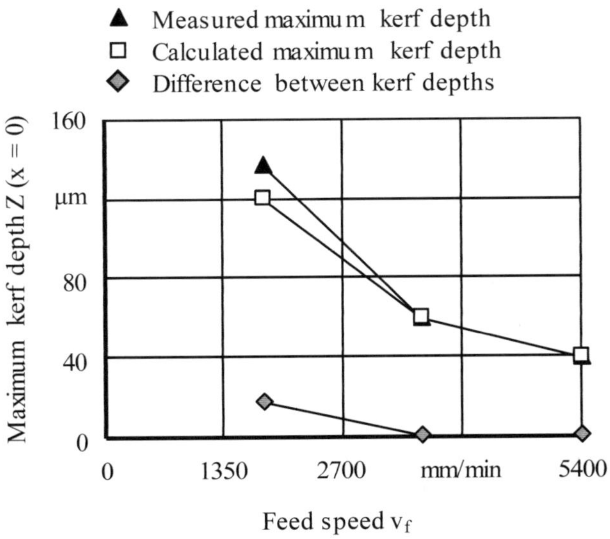

Process:
AWJ milling
Tools:
Garnet, Mesh 120, GMA
d_O = 0.25 mm
d_F = 0.76 mm
l_F = 76.20 mm
a_w = 2.00 mm
Workpiece:
TNM-B1 γ-TiAl
Process parameters:
p = 100.00 MPa
$\dot{m}_A$ = 250.00 g/min

Calculation parameters:
t_c = 0.08 ms
R = 0.40 mm
UL_x = 4.00 mm
LL_x = 0.00 mm
res_x = 0.05 mm
UL_y = 4.00 mm
LL_y = 0.00 mm
res_y = 0.05 mm
p1 = 39.90 µm
p3 = -39.90 µm

Figure 3. Calibration of the secondary etching rate $E_2(r)$

The additional material removal during low feed speeds v_f is assumed to be a result of the jet deflection or secondary jet. The secondary jet appears if a slant surface is generated during the cutting. This slant surface, around the normal vector n_{WJ} in figure 4, reflects some of the jet's energy, usually against the direction of the feed speed v_f. The reflected secondary jet can cause an additional MRR. The calculation with a primary jet energy works well only for high feed speeds v_f because the dwell time t_d is short, the kerf depth stays low and thus, the surface does not become slant. Consequently, almost no energy is reflected towards other material. The approach using the primary jet energy also works well for brittle materials. This is due to the cutting behavior of the waterjet. Bitter showed [13] that, if the angle of cut α changes from 90° to 70°, the material removal rate MRR of brittle materials decreases only about 8 %. On the other side, the same change in the angle of cut α causes an increase in the MRR of about 50 % for ductile materials. This behavior explains the higher sensitivity of ductile materials towards changes in the feed speed.

In order to establish a prediction method that includes ductile materials and low feed speeds v_f the secondary jet needs to be implemented into the approach. Hence, the direction of the secondary jet e_2 cannot be orthogonal to the workpieces surface. Figure 4 shows the fundamentals for the secondary jet calculation. In this approach it is assumed that the direction of the secondary waterjet points opposite of the surfaces normal vector n_{WJ} in the x-y-plane. In the y-z-plane the direction of the secondary jet e_2 is expected to be orthogonal to the normal vector n_{WJ} and tangential to the slant surface. The strength of the secondary etching rate is defined to close the gap between maximum measured and calculated kerf depth (i.e., figure 3). Thus, the maximum kerf depth of the secondary etching rate creates a kerf profile with $Z_2(x = 0) = 18$ µm for a feed speed of $v_f = 1800$ mm/min.

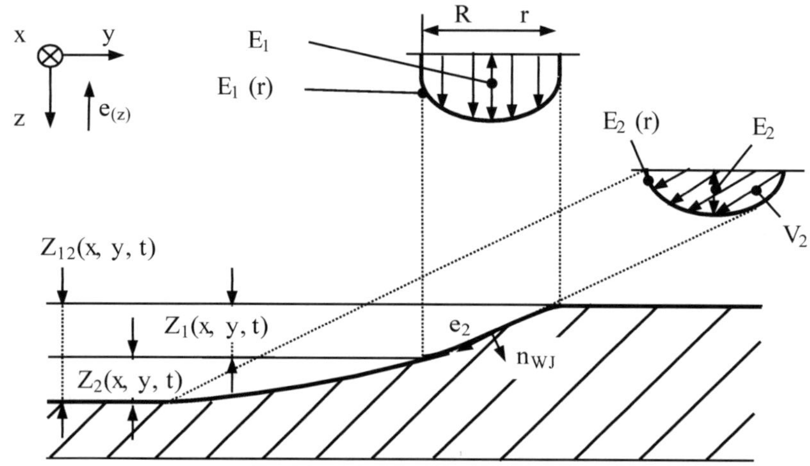

Figure 4. Fundamentals of the primary E1(r) and secondary E2(r) etching rate

In order to implement the secondary material removal Z_2, it was not only necessary to implement the secondary etching rate $E_2(r)$, but also to change the target surface itself. Since the secondary etching rate E_2 adapts the direction according to the surface and undercuts are possible, it is necessary to implement a 3D environment. The 3D environment is realized by adding a dimension for the material in z-direction. Therefore, the z-axis is defined alike the x and y axis with an upper and lower limit (UL_z, LL_z) as well as a resolution res_z. A material element M_{ijk} is added to every point in the 3D matrix. The material element M_{ijk} is defined between 0 to 1. A material element with $M_{ijk} = 1$ states an element that contains full material. The approach allows the reduction of material independent of the direction of an etching rate.

Once the 3D material matrix and the maximum secondary etching rate $E_2(r)$ were defined, an additional calculation step was implemented into the MATLAB program for every time step t after the calculation of the primary material removal. In this step, firstly, the surface normal vector n_{WJ} is calculated. Afterwards, the direction of the secondary jet e_2 is defined, and the secondary etching rate $E_2(r)$ is calculated for every point hit by the waterjet in the x-y-surface. The starting point of the secondary etching rate is the kerf profile created by the primary material removal. From this point the secondary etching rate $E_2(r)$ is directed along the direction of the secondary jet e_2. Every material element M_{ijk} that intersects with the direction of the secondary jet is checked for its value. If the material element value is $M_{ijk} = 0$, the calculations for the secondary jet continuous to the next element $M_{ijk}+e_2$. If the material element value is $M_{ijk} > 0$ the etching rate $E_2(i,j,t)$ reduces the material element M_{ijk} according to the etching rate's value.

Since the defined material element M_{ijk} exist only in a discrete manner, the direction of the secondary jet e_2 had to be limited to these discrete directions as well. Therefore, a matrix was defined to adjust the direction of the secondary jet e_2 to discrete values, with the result, that the direction of the secondary jet e_2 always points at another material element M_{ijk}. This adaption is dependent on the resolution in x, y and z. The maximum deviation between the direction calculated and the available direction is 6°.

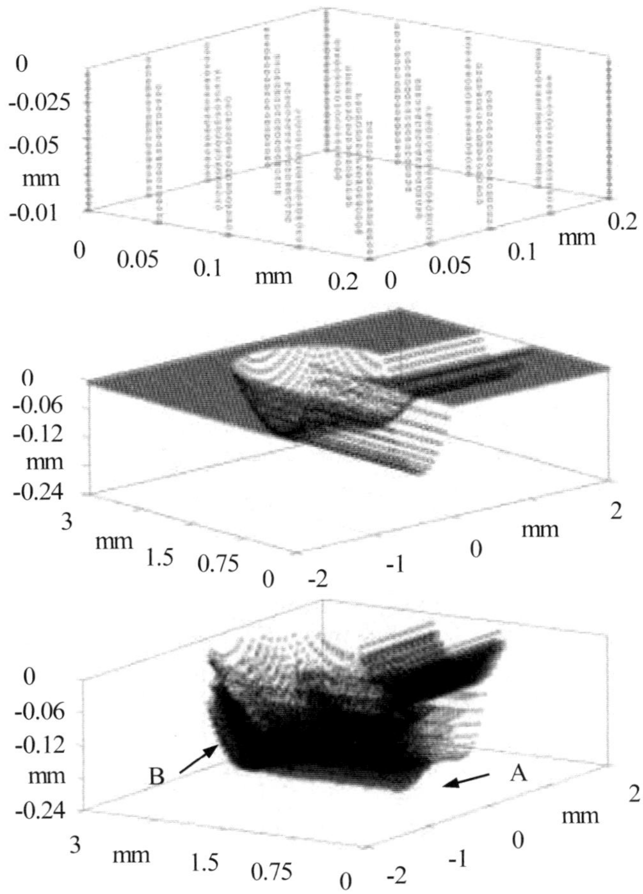

Process:
AWJ milling
Tools:
Garnet, Mesh 120, GMA
d_O = 0.25 mm
d_F = 0.76 mm
l_F = 76.20 mm
a_w = 2.00 mm
Workpiece:
TNM-B1 γ-TiAl
Process parameters:
p = 100.00 MPa
$\dot{m}_A$ = 250.00 g/min

Calculation parameters:
t_c = 0.08 ms
R = 0.40 mm
UL_x = 2.00 mm
LL_x = -2.00 mm
res_x = 0.05 mm
UL_y = 3.00 mm
LL_y = 0.00 mm
res_y = 0.05 mm
UL_z = 0.00 mm
LL_z = -0.24 mm
res_z = 5.00 µm
$\overline{p1}$ = 39.90 µm
p3 = -39.90 µm

G-Code input file

Pos.:	X	Y	v_f	R_{WJ}	G
0	3.0	1.5			
1	0.4	1.5	50		
2	0.4	2.0	30		
3	-0.4	2.0		0.4	3
4	-0.4	-1.0			

Figure 5. Results of the simulated primary E1(r) and secondary $E_2(r)$ etching rate for a free trajectory

Figure 5 shows the result of the application of the method. The depiction of the secondary jet shows a broad area of material being reached by the secondary jet. In comparison to the primary material removal, the area where the jet exits the workpiece the kerf depth decreases clearly during the last millimeter (A). This behavior is generally known from AWJ cutting [1]. Furthermore, the secondary jet seems to be stronger towards the outer side, during the cutting of a curve (B). To evaluate this effect, a section through a straight part of the trajectory and through the middle of the curve was realized and are shown in figure 6. The relations between the primary $E_1(r)$ and the secondary $E_2(r)$ kerf profile as well as their superposition $Z_{12}(r)$ for a straight trajectory are shown on the left side of the figure. For the straight trajectory the diagram states the relations given in figure 3, where it has been calibrated.

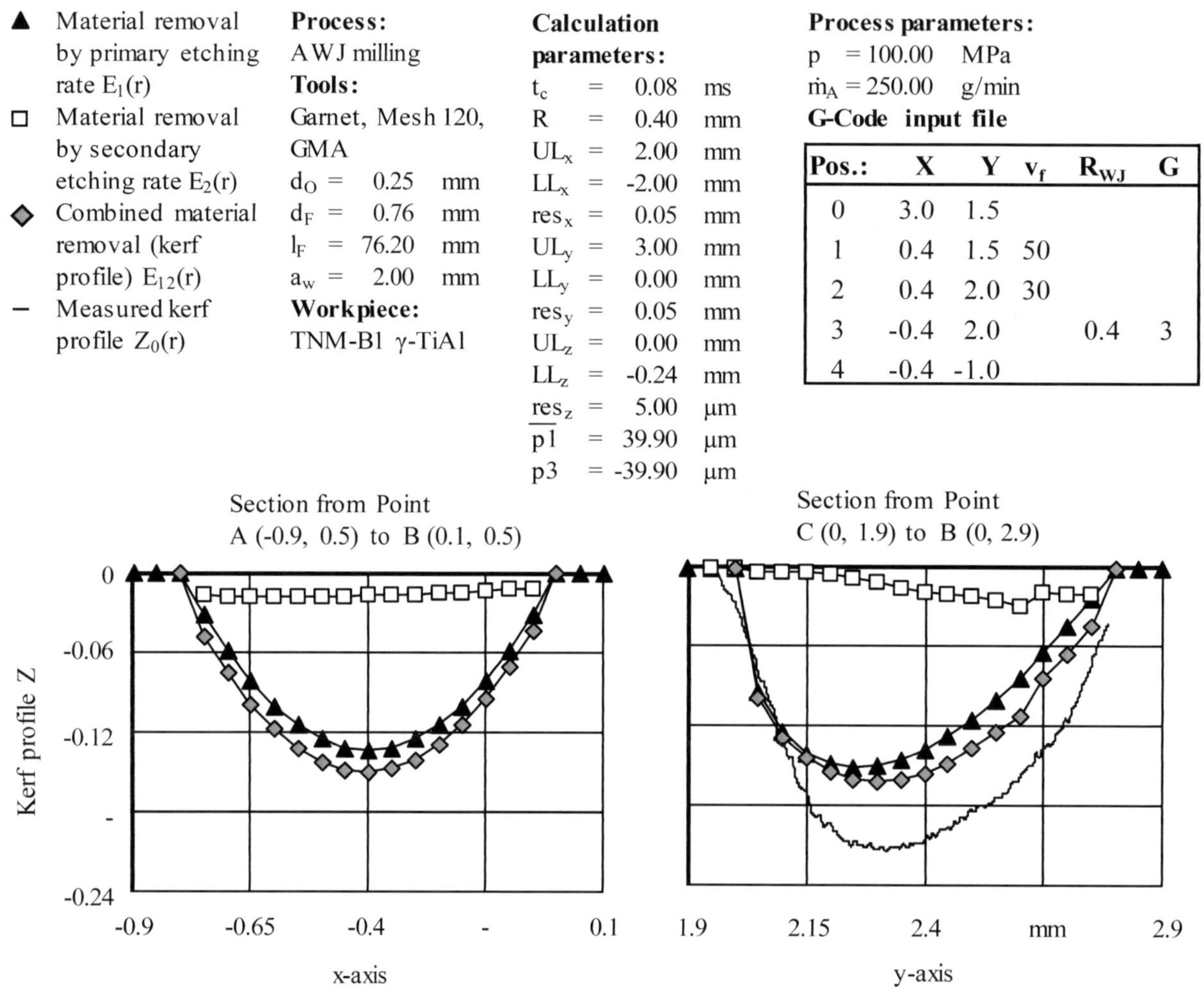

Figure 6. Comparison of kerf profiles Z

On the right side of the figure 6 the same profiles are shown for a curve. The diagram shows how the relations and the distribution of the calculated kerf profiles change. First, the peak of primary kerf profile $Z_1(r)$ is displaced from the middle of the jet towards the inner side (left side). The effect of the secondary material removal Z_2 grows stronger towards the outer side of the radius (right side). In combination a calculated kerf profile Z_{12} similar to the one of straight trajectories ensues. The effects of cutting curves have previously been investigated. This investigation states that the overall material removal does not change much from straight trajectories [15]. Thus, the calculated results seem to be in accordance with these findings. The measured kerf profile is about a third deeper than calculated. Hence, a direct validation of the calculation on the measured kerf profile Z_0 is not possible. The difference between the measured and the calculated profile can be explained by the drop in the feed speed v_f due to the inevitable deceleration and acceleration of the manufacturing machine. The problem is described in detail by Klocke [10]. If the differences between the depth of the profiles are left aside, the form of both the analytical and the measured profiles can still be compared. For this comparison, first, the coefficient of determination between the measured kerf profile $Z_0(r)$ and the primary kerf profile $Z_1(r)$ is calculated with $R^2 = 76$ %. If the secondary etching rate E_2 is added to the primary profile Z_l the coefficient of determination becomes $R^2 = 86$ %. Thus, the test result shows that applying the secondary jet energy improves the quality for the prediction for lower feed speeds v_f on ductile material. The model will be particularly helpful to design the cutting strategies, with minimized errors, for geometries given in figure 7.

a) b)

Figure 7. a) Examples of geometries suitable for modelling with the primary and secondary jet energy; b) Manufacturing of various angle of cuts α

3. CONCLUSION

A model to predict the kerf profile for AWJ milling by implementing a primary and secondary jet energy has been introduced and the capabilities of the approach have been demonstrated. The model is based on the energy of the waterjet as introduced by Axinte [6], broadened by applying a secondary jet and its energy. This extension allows a more accurate prediction of the kerf profiles for lower feed speeds and in particular for ductile materials. Furthermore, the jet movement can be controlled by a G-Code file allowing a wide range of movements. In addition, the target material surface has been set up to enable undercuts in the material. The results show that the calculation can reproduce specific behaviors of the waterjet such as the exiting behavior and the kerf formation cutting radii.

4. OUTLOOK

As described, ductile materials react very sensitively towards changes in the angle of cut α. Therefore, the primary and the secondary jet energy must be calibrated and implemented in the model for different angles of cut $\alpha \neq 90°$ (figure 7). In addition, the calculation results might be further improved by applying a normal vector for each finite element n_{WJij} that is targeted by the waterjet. The result would be a different secondary jet direction for each finite waterjet. Once the comprehensive model has been established, it should be able to, not only predict the kerf profile, but also be used to quickly apply changes in the waterjet parameter settings for specific problems and analyze the resulting kerf profile Z(r). These findings could be used to derive acceleration and accuracy requirements for new AWJ milling machines.

5. ACKNOWLEDGMENTS

This paper is based on results acquired in the project DFG UH 100/206-2 and DFG UH 100/165-3, which is kindly supported by the DEUTSCHE FORSCHUNGSGEMEINSCHAFT (DFG).

6. REFERENCES

[1] Hashish M.: A Modeling Study of Metal Cutting with Abrasive Waterjets. Journal of Engineering Materials Technololy 106 (1986), p. 88 - 100.

[2] Hashish, M.: Wear in Abrasive-Waterjet Cutting Systems. Wear of Materials (1987), p. 769 - 776.

[3] Zeng, Jiyue; Kim, Thomas J.: An erosion model for abrasive waterjet milling of polycrystalline ceramics. Wear 199 (1996), p. 275 - 282.

[4] Hashish, M.: Controlled-Depth Milling of Isogrid Structures with AWJs. Journal of Manufacturing Science and Engineering. 120 (1998), p. 21 - 27.

[5] Fowler, G.; Shipway, P; Pashby, I.: Abrasive water-jet controlled depth milling of Ti6Al4V alloy - an investigation of the role of jet - workpiece traverse speed and abrasive grit size on the characteristics of the milled material. Journal of Materials Processing Technology (Journal of Materials Processing Technology). 161 (2005) p. 407 - 414.

[6] Axinte DA, Srinivasu DS, Billingham J, Cooper M. Geometrical modelling of abrasive waterjet footprints: A study for 90° jet impact angle. CIRP Annals - Manufacturing Technology 59, 2010, p. 341 - 346.

[7] Faltin, F.: Endkonturnahe Schruppbearbeitung von Titanaluminid mittels Wasserabrasivstrahlen mit kontrollierter Schnitttiefe. Berichte aus dem Produktionstechnischen Zentrum Berlin. Hrsg.: Uhlmann, E. Dissertation, Technische Universität Berlin. Stuttgart: Fraunhofer IRB, 2018.

[8] ElTobgy, M.; Ng, E-G.; Elbestawi, M. A.: Modelling of Abrasive Waterjet Machining, A New Approach. CIRP Annals - Manufacturing Technology 54, 2005, p. 285 - 288.

[9] Van Bui H, Gilles P, Sultan T, Cohen G, Rubio W. A new cutting depth model with rapid calibration in abrasive water jet machining of titanium alloy. The International Journal of Advanced Manufacturing Technology 93 (2017), p. 1499 - 1512.

[10] Klocke F, Schreiner T, Schüler M, Zeis M. Material Removal Simulation for Abrasive Water Jet Milling. Procedia CIRP 68, 2018, p. 541 - 546.

[11] Kong MC, Anwar S, Billingham J, Axinte DA. Mathematical modelling of abrasive waterjet footprints for arbitrarily moving jets: Part I-single straight paths. International Journal of Machine Tools and Manufacture 53 (2012), p. 58 - 68.

[12] Finnie, I.: The mechanism of erosion of ductile metals. Proceedings of the 3rd US National Congress of Applied Mechanics, Rhode Island, USA, 1958, p. 572 - 532.

[13] Bitter, J. G. A.: A study of erosion phenomena - Part II. Wear, Vol. 6, Issue 3, 1963, p. 169 - 190.

[14] Laurinat, A.: Abtragen mit Wasserabrasivinjektorstrahlen. Dissertation, Universität Hannover, 1994.

[15] Uhlmann, E.; Männel, C.: Effects of Abrasive Waterjet Trepanning on the Kerf Formation. Proceedings of the 24th BHR: BHR Group, 2018, p. 155 - 165.

7. NOMENCLATURE

Symbol	Unit	Definition
α	°	Angle of cut
AWJ	-	Abrasive waterjet
C	1	Constante
d	mm	Diameter
d_c	mm	Depth of cut
d_d	mm	Orifice diameter
d_f	mm	Focus nozzle diameter
DFG	-	German Research Foundation
E	mm/s	Etching rate
e_2	1	Direction of the secondary jet
G	-	G-Code for the direction of the waterjet
LL	mm	Lower limit
l_f	mm	Focus nozzle length
$\dot{m}_A$	g/min	Abrasive flow rate
n	1	Normal vector
n_{WJ}	1	Surfaces normal vector
n_{WJij}	1	Surfaces normal vector for particular position
p	MPa	Pressure
P_{WJ}	mm	Position of the waterjet
R	mm	Maximum radius of the waterjet
R_{WJ}	mm	Radius of the waterjet's trajectory
r	mm	Radius
res	mm	Resolution
t	S	Time step
t_c	S	Control time
t_d	S	Dwell time
UL	mm	Upper limit
V	m/s	Waterjet velocity
v_f	mm/min	Feed speed
V(r)	m/s	Waterjet velocity profile
q	0	Power factor
Z	mm	Kerf profile

Indices	Definition
0	Measured
1	Primary
2	Secondary
i	X-direction
j	Y-direction
k	Z-direction
$\overline{\ }$	Normalized value

2019 WJTA Conference and Expo
November 11-13 ● New Orleans, Louisiana

Paper

ANALYSIS OF THE EFFECT OF WATER JET PARAMETERS ON THE PERFORMANCE DURING PREPARATION OF CUTTING TOOL EDGE BY FINE ABRASIVE WATER JET

Yuyong Lei, Yuqiang Wu, Wei Yang
Xihua University
Chengdu, P. R. China

Ying Zhang
Tennessee Tech University
Cookeville, Tennessee, USA

ABSTRACT

Preparation of cutting tool edge using fine abrasive water jet (FAWJ) was proposed. A specific designed water jet machine tool with four axes was developed. Experiments on preparation of cutting edge of carbide chasers for petroleum steel pipes using FAWJ have been carried out. Taguchi test method was applied in design of experiments. The effects of process parameters on the radius of cutting edge as well as the surface roughness of the cutting tool were investigated and analyzed. Based on the range analysis, among the five process parameters selected, the traverse velocity of the FAWJ exhibits the greatest significance on the prepared cutting edge radius, followed by abrasive flow rate, size of abrasive, pressure and stand-off distance. Nevertheless the most significant impact factor of the designed parameters on the prepared surface roughness is abrasive flow rate, followed by the traverse velocity, stand-off distance, size of abrasive and water pressure. From the analysis of variance, the abrasive flow rate has the most significant impact on both the prepared cutting edge radius and the surface roughness. The traverse velocity also has important impact on the prepared cutting edge radius and the surface roughness.

Organized and Sponsored by WJTA®

1. INTRODUCTION

Cutting tool plays an important role in the machining process. Carbide chaser is an important tool for manufacturing thread of the petroleum steel pipes. Generally a carbide chaser has more than one tooth, as shown in Figure 1. The cutting teeth of the carbide chaser play a significant role in the machining process. In order to improve cutting efficiency, extend tool life and therefore reduce manufacturing costs, a lot of work has to be done . This includes optimization of structure of cutting tool, modification of the tool edge geometry, development of new tool materials as well as tool coating technology[1-5]. The early research shows that most of the defects on the cutting tool, such as micro cracks, burring, chipping etc. can be eliminated by preparation procedure[6]. Therefore researchers have paid enough attention to cutting tool preparation.

The picture of a tooth on the carbide chaser under measuring microscope 107JPC is shown in Figure 2. It can be seen that the cutting edge is very sharp and zigzag burring exists on the cutting edge. On the other hand, linear waviness appears on the surface of the carbide chaser.

The purpose of cutting edge preparation is to strengthen the cutting edge or prepare a surface for deposition of hard coatings. There are three types of cutting edge for cutting tool, i.e. sharp edge, honed or round edge and T-land/chamfer edge in most commercial applications [5], as shown in Figure 3. A chamfer edge combined with additional hone is also available. Round edges are usually employed in finish cuts and it can reduce the initiation of notch wear. Chamfer edges are used in heavy chip loads and interrupted cuts because it can further strengthen tool edge. Thereafter the working life of cutting tool can be greatly improved.

Preparation of cutting edge using abrasive water jet (AWJ) was proposed in our early research. The experimental study has shown the feasibility of preparation of the cutting edge of carbide inserts using AWJ [6]. The zigzag burring existing on the cutting edge was successfully eliminated through preparation by AWJ. Also the very sharp cutting edge was rounded after preparation by AWJ compared with the original one. Thus the cutting edge strength can be enhanced during this course. The results show that preparation of cutting edge of the tool by AWJ is efficient, environmental friendly and higher quality compared with other preparation method. The further experimental study on the preparation of cutting edge using FAWJ is necessary and the effects of experimental parameters on the radius of cutting tool edge as well as surface quality were analyzed in this paper.

2. EXPERIMENTAL SETUP

The experimental setup for preparation of cutting edge of carbide chaser was based on HJ300 abrasive water jet machine, as shown in Figure 4. The AWJ machine has four axes, i.e. linear axis X/Y/Z and rotary axis A. The X, Y, and Z axes are numerically controlled. The abrasive nozzle assembly was driven by X/Y/Z axies. Therefore the abrasive nozzle can map any 2D shape on the machine table.

The rotary (A) axis is used to hold the fixture for carbide chaser and adjust the carbide chaser to perpendicular to the abrasive water jet. Because the fine abrasives were used, it is difficult for the fine abrasives to entrain into the nozzle. So an auger mechanism was designed and used to feed the fine abrasives. The flow rate of abrasives can be precisely controlled by regulating the rotating speed of the electrical motor of the auger mechanism[7], as shown in Figure 5.

In order to prepare of cutting edge of carbide chaser for petroleum steel pipes, a specially designed abrasive nozzle (focusing tube) with rectangular inside profile was designed, as shown in Figure 6. The inside profile of abrasive nozzle was composed of three sections, i.e., the contraction section, the cylinder section and the diffusion section. The detailed parameters are showed in Table 1.

In the experiments, carbide chaser for the production of petroleum steel pipes was used as a specimen, as shown in Figure1b. Because of its relatively small size, specially designed fixture for carbide chaser was used during cutting edge preparation with FAWJ. The detailed experimental parameters are listed in Table 2.

3. MEASUREMENTS

An optical 3D measuring device for the cutting tool (MikroCAD) from GFM was used for measuring the radius of prepared cutting edge of carbide chaser, as shown in Figure 7. MikroCAD is a customized system for automated measurement of cutting edge rounding. The profile of the cutting edge of the carbide chaser can be easily scanned by a narrow laser. Equipped with specialized software, it has image processing, data acquisition and data analyzing functions. Therefore the rounding radius of the cutting edge along carbide chaser can be reported by MikroCAD system.

4. EXPERIMENTAL RESULTS AND ANALYSIS

More than 100 thread carbide chasers were prepared in experiments. The prepared samples of carbide chasers are as shown in Figure 8. One of the tooth's camera images of the carbide chaser after preparation with FAWJ was shown in Figure 9a. And the rounding radius for specific section between the blue lines after preparation with fine abrasive water jet was shown in Figure 9b.

Figure10 depicts the variation of rounded radius of the cutting edge along the one tooth of the carbide chaser after preparation by FAWJ. As shown in Figure10, the rounded radius is uneven along the cutting tooth of the carbide chaser, although the variation of the radius is limited. The detailed data of radius after preparation are listed in Table 3.

The comparison of radius of the cutting edge before and after preparation is shown in Figure11. It can be seen from the profile of cutting edge that there are significant changes compared with the original tool edge. The very sharp cutting edge was rounded after preparation by fine abrasive water jet.

In order to explore the performance of preparation of cutting tool edge by fine abrasive water jet and find the effects of process parameters on the radius of cutting edge as well as the surface roughness of the cutting tool, detailed experiments are needed.

4.1 Design of Detailed Experiments

Taguchi test method was applied in design of experiments. According to the preliminary results on preparation of the cutting tool edge, three levels and five factors were considered, as shown in Table 4. The interactive influence of five independent factors was neglected, therefore an final test design of $L_{27}(3^5)$ was applied in this work, as shown in Table 4. The rounded radius of the cutting tool edge and the surface roughness of the cutting tool were considered as the performance during preparation by FAWJ. The results are presented in Table 5.

4.2 Analysis of the Effects of Process Parameters on the Radius of Cutting Edge

In order to enhance the strength of cutting tool edge, the rounded radius of the cutting edge of carbide chaser in the range of 40-70μm is expected. Based on our experiments, the rounded radius is a little bit smaller. From this point, higher water pressure or coarser abrasive should be used during preparation. It should be pointed out that coarser abrasive will be result in higher surface roughness on the rake face and the flank face. Therefore two steps of preparation of cutting edge of carbide chaser for petroleum steel pipes maybe suggested.

4.2.1 The Range Analysis of the Prepared Cutting Edge Radius

The range analysis was used for the analysis of prepared cutting edge radius based on Taguchi experiments in Table 5. The range analysis results for prepared cutting edge radius obtained was shown in Table 6.

In Table 6, K_i is the sum of experimental results r_i based on Taguchi Table 5 when process parameter is at level i. A_i is the average of K_i and R is max-min difference among A_i. Thus they are given in equations (1), (2) & (3).

$$K_i = \sum_{i=1}^{9} r_i \tag{1}$$

$$A_i = \frac{1}{9}\sum_{i=1}^{9} r_i \tag{2}$$

$$R = max(A_i) - min(A_i) \tag{3}$$

Based on the range analysis for prepared cutting edge radius in Table 6, the significant impact factor of design parameters on the prepared cutting edge radius was distinguished and the optimal combination of design parameters in test conditions was found. As shown in Table 6, among the five process parameters, the traverse velocity of the FAWJ exhibits the greatest significance, followed by abrasive flow rate, size of abrasive, water pressure and stand-off distance. The results suggest that the cutting edge radius during preparation is most sensitive to the traverse velocity of the FAWJ and abrasive flow rate.

The influence of designed process parameters on the prepared cutting edge radius based on the sum of experimental results K_i was depicted in Figure12. Apparently a larger rounded radius on the cutting edge of the carbide chaser is expected. Therefore the optimal combination of designed parameters under experimental conditions was determined, i.e., $p = 120$ Mpa, $u = 100$ mm/min, $m_a = 1.0$ g/s, $s = 240\#$, $s_d = 12mm$, as shown in Figure12.

4.2.2 Analysis of Variance of the Prepared Cutting Edge Radius

Analysis of variance (ANOVA) is a statistically based, objective decision–making tool for detecting any differences in average performance of groups of parameters tested. It helps formally test the significance of all main factors and their interactions by comparing the mean square against an estimate of the experimental errors at specific confidence levels.

With the help of ANOVA, the influence of experimental parameters on the cutting edge radius could be obtained. The results for the cutting edge radius by ANOVA based on experimental data in Table 5 were represented in Table 7.

The F ratio is generally used to determine the significance of the factor effects, which is defined as the ratio of the mean square and the error mean square. From Table 7, at 95% confidence level, the experiment parameter that has the greatest F value (2.08) is abrasive flow rate. It indicates that the abrasive flow rate has most significant impact on the prepared cutting edge radius, while the traverse velocity and the size of abrasive also has the important impact on the prepared cutting edge radius. The significance of water pressure and stand-off distance is relatively small. The water pressure has less impact on the prepared cutting edge radius may be due to the less difference among the three levels.

4.2.3 Regression Analysis of the Prepared Cutting Edge Radius

Regression analysis can determine the relationship between dependent and independent variables. Here it was used to predict the prepared cutting edge radius as a function of the process parameters of preparation of cutting tool edge by FAWJ, i.e., water pressure, traverse velocity, abrasive flow rate, stand-off distance and the size of abrasive.

In this analysis, the prepared cutting edge radius was considered to be a response (output), while parameters such as water pressure, traverse velocity, abrasive flow rate, stand-off distance and the size of abrasive were considered to be independent (input) variables[10]. Generally there is a nonlinear relationship for the multi-factor model, the nonlinear function of the prepared cutting edge radius was given in equation 4.

$$r = a_0 p^{a_1} u^{a_2} m_a^{a_3} S^{a_4} s_d^{a_5} \tag{4}$$

Where a_0 is a constant coefficient; a_1, a_2, a_3, a_4,and a_5 are undetermined coefficient.

With the help of MATLAB toolbox, nlinfit function was used to find the undetermined coefficient based on the Taguchi experiment results shown in Table 5. The regression analysis results obtained are shown in Table 8.

Therefore the regression equation of the prepared cutting edge radius r on the carbide chaser was expressed as follows.

$$r = 26.0676 p^{0.0309} u^{0.1300} m_a^{-0.2023} S^{-0.1985} s_d^{0.0934} \quad (5)$$

In order to verify the regression analysis of he prepared cutting edge radius, 6 set of experimental data were selected. The relative error between the experimental values and predict values based on the regression equation(5) are around 5.189%-7.5897%.

4.3 Analysis of the Effects of Process Parameters on Tool Surface Roughness

4.3.1 The Range Analysis of the Tool Surface Roughness

The range analysis results for prepared tool surface roughness obtained was shown in Table 9. Based on the range analysis results in Table 9, the significant impact factor of design parameters on the prepared surface roughness is abrasive flow rate, followed by the traverse velocity, stand-off distance, size of abrasive particles and water pressure. In order to achieve better surface finish on the carbide chaser after preparation, The better the smaller of the surface roughness R_a is. Therefore the optimal combination of designed parameters under experimental conditions is as follows, i.e., $p = 80$ Mpa, $u = 80$ mm/min, $m_a = 2.0$ g/s, $s = 400\#$, $s_d = 9mm$, as shown in Figure13.

4.3.2 Analysis of Variance of the Tool Surface Roughness

Analysis of Variance of the prepared tool surface roughness based on Taguchi experimental data in Table 5 were represented in Table 10.

As shown in Table 10, the experiment paremeter that has the greatest F value (21.31) is abrasive flow rate. It indicates that the abrasive flow rate has most significant impact on the surface roughness during the preparation of cutting tool by FAWJ, while the significance of traverse velocity, stand-off distance and the water pressure is relatively small. The interesting result is that the size of abrasive has the least impact on the surface roughness. This is probably due to the small difference among the three types of abrasive size. Thus the surface roughness of the cutting tool is more sensible to the changes in the abrasive flow rate than that of abrasive size.

4.3.3 Regression Analysis of the Prepared Surface Roughness of Carbide Chaser

Generally there is a nonlinear relationship for the multi-factor model, the nonlinear function of the cutting tool surface roughness was given in equation 6.

$$Ra = c_0 p^{c_1} u^{c_2} m_a^{c_3} S^{c_4} s_d^{c_5} \quad (6)$$

Where c_0 is a constant coefficient; c_1, c_2, c_3, c_4, and c_5 are undetermined coefficient.

With the help of MATLAB toolbox, nlinfit function was used to find the undetermined coefficient based on the Taguchi experiment results shown in Table 5. The regression analysis results for coefficient obtained are shown in Table 11. The regression equation of the surface roughness R_a on the prepared carbide chaser is as follows.

$$Ra = 0.0106 p^{0.2319} u^{0.4319} m_a^{-0.8371} s^{0.0039} s_d^{0.5809} \quad (7)$$

According to the experimental data of verification, the maximum relative error of the regression equation(7) is 7.55%.

5. CONCLUSIONS

Experiments on preparation of cutting edge of carbide chaser for petroleum steel pipes using FAWJ have been carried out. The experimental results show that the micro defects on the cutting tool surface, such as micro cracks, waveness, burring, chipping etc. were elimilated. The very sharp edge with irregular zigzag shape was rounded to a radius around 20-30μm by FAWJ. Based on the analysis from the Taguchi experiments, the following conclusions can be drawn.

1) From the range analysis, among the five process parameters selected, the traverse rate of the FAWJ exhibits the greatest significance on the prepared cutting edge radius, followed by abrasive flow rate, size of abrasive, pressure and stand-off distance. Nevertheless the significant impact factor of design parameters on the prepared surface roughness is abrasive flow rate, followed by the traverse velocity, stand-off distance, size of abrasive and water pressure.

2) Based on analysis of variance, the abrasive flow rate has the most significant impact on both the prepared cutting edge radius and the surface roughness. The traverse velocity also has important impact on the prepared cutting edge radius and the surface roughness.

3) Under experimental conditions, the nonlinear relationship between the prepared cutting edge radius and the process parameters was given by $r = 26.0676 p^{0.0309} u^{0.1300} m_a^{-0.2023} S^{-0.1985} s_d^{0.0934}$. The relative error between the experimental values and predict values are around 5.189%-7.5897%.

4) Under experimental conditions, the regression equation for the surface roughness was determined by $Ra = 0.0106 p^{0.2319} u^{0.4319} m_a^{-0.8371} s^{0.0039} s_d^{0.5809}$. According to the experimental data of verification, the maximum relative error for the regression equation is 7.55%.

6. ACKNOWLEDGEMENT

The authors would like to thank the support of project (18ZA0451) financed by Educational Bureau of Sichuan Province, China and the Key Subject Project (XK-JX-2018-2) financed by Xihua University, People's Republic of China.

7. REFERENCES

[1] D. Bouzakis et al. Optimization of the cutting edge roundness and its manufacturing procedures of cemented carbide inserts, to improve their milling performance after a PVD coating deposition [J]. Surface and Coatings Technology,163-164(2003):625-630.
[2] J.Rech, Y. C. Yen, M.J. Schaff, et al. Influence of cutting edge radius on the wear resistance of PM-HSS milling inserts[J].Wear , 2005(259):1168-1176.
[3] K. Shintani, M. Ueki, Y. Fujimura. Optimum tool geometry of CBN tool for continuous turning of carburized steel [J]. Interactional Journal of Machine Tools and Manufacture, 1989, 29(3):403-413.
[4] Gui Yupeng, et al. Discussion of technology of cutting tool edge preparation [J]. Machinist Metal Cutting, 2004(6): 43-44.
[5] Yung-Chang Yen, Anurag Jain, Taylan Altan. A finite element analysis of orthogonal machining using different tool edge geometries [J]. Journal of Materials Processing Teehnology, 146(2004):72-81.
[6] Yang Zhen, Lei Yuyong, Zou Hu, Lv Hua, Jia Weiguang , Experiments on Preparation of Cutting Tool Edge Using Abrasive Water Jet , Advanced Materials Research, v 337, p 146-150.
[7] Dai Liangbo, Lei Yuyong, et al. Installation for Precise Feed Fine abrasive and Its Application in Abrasive Water Jet [J]. Manufacturing Technology & Machine Tool, 2010(9):17-25.
[8] Wan Qingfeng , Lei Yuyong , Liu Kefu , Chen Zhongmin , Pan Zhengzheng, Research on model of tool edge preparation based on fine abrasive water jet, Modern Manufacturing Engineering [J], 2013 , No.9:95-99
[9] Wan Qing-feng , Lei Yu-yong , Chen Zhong-min , Pan Zheng-Zheng, Research on regression Model of Tool Preparation Surface Quality Based on Fine abrasive Water Jet, Modular Machine Tool & Automatic Manufacturing Technique, 2014 , No.1:31-34
[10] Chen Xuan, Lei Yuyong, et al., Regression Analysis of Tool Preparation Tool Edge Radius Based on Fine abrasive Water Jet, Machine Tool & Hydraulics, 2016, Vol.44(1):96-98

8. NOMENCLATURE

d - Mixing tube inside diameter /mm
D - Mixing tube outside diameter /mm
a,b -The lenth and width of mixing tube outlet
L_1, L_2, L -The lenth of entrance segment, cylinder segment and whole respectively
p - Water jet pressure/ *Mpa*
u - Traverse velocity/ mm/min
m_a - Abrasive flow rate/ g/min
s - Abrasive particle size/ mesh
s_d - Stand-off distance/mm
r_0 , r - the cutting edge radius before ande after prepared respectively

R_a - surface roughness after preparation

K_i - the sum of experimental results r_i when process parameter is at level i

A_i - the average of K_i

R - max-min difference among the A_i

a_0, c_0 -constant coefficient

a_1, a_2, a_3, a_4, a_5 -undetermined coefficient

c_1, c_2, c_3, c_4, c_5 - undetermined coefficient

Table 1. Parameters of Rectangular Abrasive Nozzle

α_1(°)	60	d(mm)	0.15
L_1(mm)	5.75	D(mm)	6.35
L_2(mm)	15	axb(mm)	2.0×0.5
L(mm)	50.8	α_2(°)	30

Table 2. Experimental Parameters for FAWJ

Water Jet Pressure p **/ MPa**	120
Stand-off Distance s_d **/mm**	15
Abrasive Flow Rate m_a **/ g/min**	90.61
Abrasive Type	SiC
Abrasive Particle Size s **/ mesh**	240-400
Traverse Velocity u **/ mm/min**	80-160

Table 3. Measured Radius of the Cutting Edge

Radius	**Unit**	**Min**	**Max**	**Mean**	**Standard Deviation**
r	μm	24.4	28.1	25.9	0.79

Table 4. Taguchi Test Design

Factors / **Level**	p/MP_a	$u/mm/\text{min}$	$m_a/g/s$	$s/mesh$	s_d/mm
1	120	80	1.0	240	9
2	100	120	1.5	325	12
3	80	160	2.0	400	15

Table 5. Prepared Results by FAWJ

Test No.	p/MP_a	$u/mm/\min$	$m_a/g/s$	$s/mesh$	s_d/mm	$r/\mu m$	$R_a/\mu m$
1	1	1	1	1	1	24.5	0.821
2	1	1	1	1	2	22.9	0.837
3	1	1	1	1	3	23.7	0.907
4	1	2	2	2	1	18.4	0.812
5	1	2	2	2	2	22.1	0.925
6	1	2	2	2	3	23.5	1.397
7	1	3	3	3	1	21.1	0.556
8	1	3	3	3	2	20.4	0.563
9	1	3	3	3	3	20.5	0.554
10	2	1	2	3	1	19.3	0.404
11	2	1	2	3	2	17.1	0.492
12	2	1	2	3	3	16.0	0.843
13	2	2	3	1	1	24.9	0.748
14	2	2	3	1	2	21.2	0.595
15	2	2	3	1	3	19.8	0.556
16	2	3	1	2	1	16.4	0.944
17	2	3	1	2	2	28.5	1.001
18	2	3	1	2	3	17.9	1.418
19	3	1	3	2	1	15.3	0.446
20	3	1	3	2	2	14.2	0.514
21	3	1	3	2	3	22.8	0.294
22	3	2	1	3	1	20.9	0.924
23	3	2	1	3	2	25.7	0.941
24	3	2	1	3	3	26.4	1.307
25	3	3	2	1	1	22.8	0.740
26	3	3	2	1	2	23.3	0.756
27	3	3	2	1	3	21.9	0.951

Table 6. Range Analysis for the Prepared Cutting Edge Radius by Taguchi

	$r/\mu m$				
	p	u	m_a	s	s_d
K_1	197.10	175.80	206.90	205	183.60
K_2	181.10	202.90	184.40	179.10	195.40
K_3	193.30	192.80	180.20	187.40	183.50
A_1	21.90	19.53	22.99	22.78	20.40
A_2	20.12	22.54	20.49	19.90	21.71
A_3	21.48	21.42	20.02	20.82	21.38
R	1.778	3.011	2.967	2.878	1.311
the Order of Significance of the Factors: $u \rightarrow m_a \rightarrow s \rightarrow p \rightarrow s_d$					

Table 7. Analysis of Variance of the Prepared Cutting Edge Radius

Variance Source	Degrees of Freedom (DF)	Sum of Square (SS)	Mean Square (MS)	F Value (F)	P Value (P)
p/MPa	2	15.529	7.764	0.7	0.509
u/mm/min	2	4.682	20.841	1.89	0.183
m_a/g/s	2	45.807	22.903	2.08	0.157
S/mesh	2	38.869	19.434	1.76	0.203
S_d/mm	2	8.402	4.201	0.38	0.689
Error	16	176.331	11.020		
Total	26	326.62			

Table 8. Regression Analysis of Prepared Cutting Edge Radius

a_0	a_1	a_2	a_3	a_4	a_5
26.0676	0.0309	0.1300	-0.2023	-0.1985	0.0934

Table 9. Range Analysis for the Prepared Surface Roughness by Taguchi

	$R_a / \mu m$				
	p	u	m_a	s	s_d
K_1	7.372	5.558	9.1	6.911	6.395
K_2	7.001	8.205	7.32	7.751	6.624
K_3	6.873	7.483	4.826	6.584	8.227
A_1	0.8191	0.6175	1.011	0.7678	0.7105
A_2	0.7778	0.9116	0.8133	0.8612	0.736
A_3	0.7636	0.8314	0.5362	0.7315	0.9141
R	0.0554	0.2941	0.4748	0.1296	0.2035
the Order of Significance of the Factors: $m_a \rightarrow u \rightarrow s_d \rightarrow s \rightarrow p$					

Table 10. Analysis of Variance of the Tool Surface Roughness

Variance Source	Degrees of Freedom (DF)	Sum of Square (SS)	Mean Square (MS)	F Value (F)	P Value (P)
p/MPa	2	0.081	0.0403	1.68	0.218
u/mm/min	2	0.416	0.2080	8.66	0.003
m_a/g/s	2	1.024	0.5121	21.31	0
S/mesh	2	0.015	0.0075	0.31	0.737
S_d/mm	2	0.221	0.1107	4.61	0.026
Error	16	0.384	0.0240		
Total	26	2.142			

Table 11. Regression Analysis of Prepared Surface Roughness

c_0	c_1	c_2	c_3	c_4	c_5
0.0106	0.2319	0.4319	-0.8371	0.0039	0.5809

a) Carbide Chaser in Production Thread b) Carbide Chaser

Figure 1. Carbide Chaser for Manufacturing Thread of a Petroleum Steel Pipe

Figure 2. Picture of One Tooth of Carbide Chaser (Magnification of 40×400)

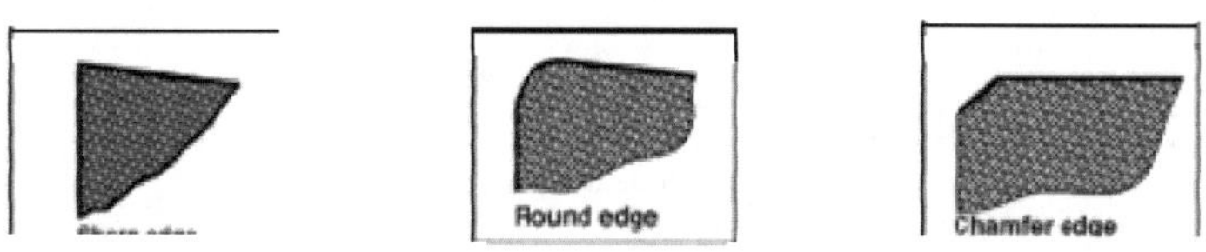

a) Sharp edge b) Round edge c) Chamfer edge

Figure 3. Typical Types of Cutting Edge for Cutting Tool

Figure 4. Experimental Setup for Cutting Edge Preparation

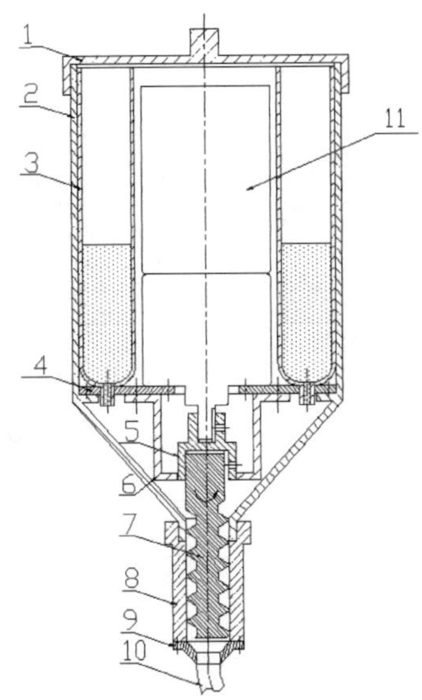

Figure 5. Fine Abrasives Precise Feed System
1-cover 2-shell 3-cavity 4-Support 5-coupling
6- blocking tube 7-auger 8-sheathing 9-joint
10-hose 11-geared motor

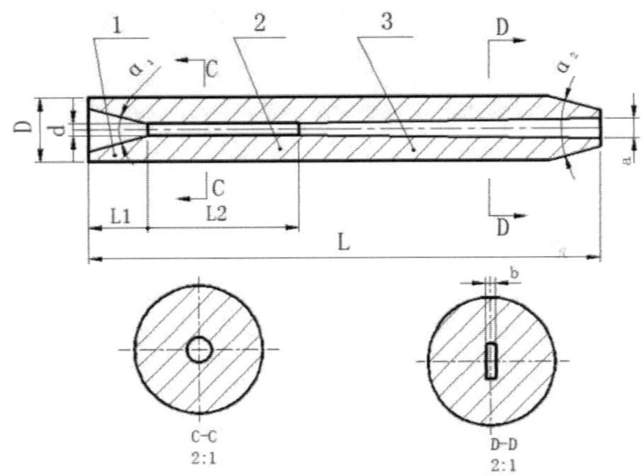

Figure 6. Rectangular Abrasive Nozzle

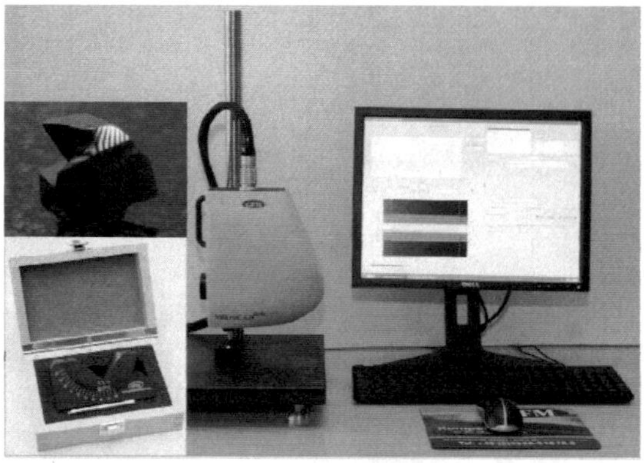

Figure 7. MikroCAD System

Figure 8. Prepared Samples

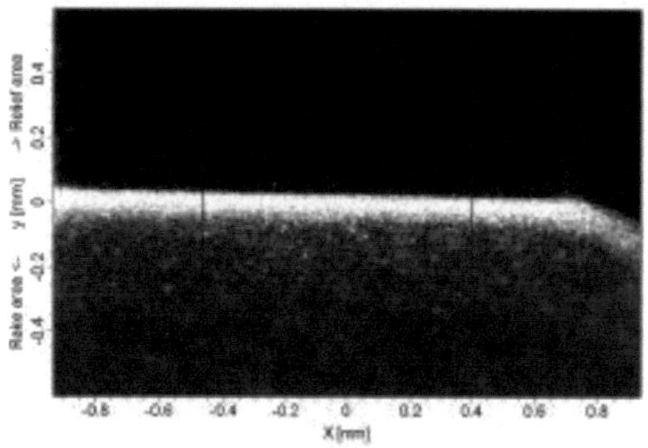

a) Camera Image of the Cutting Edge

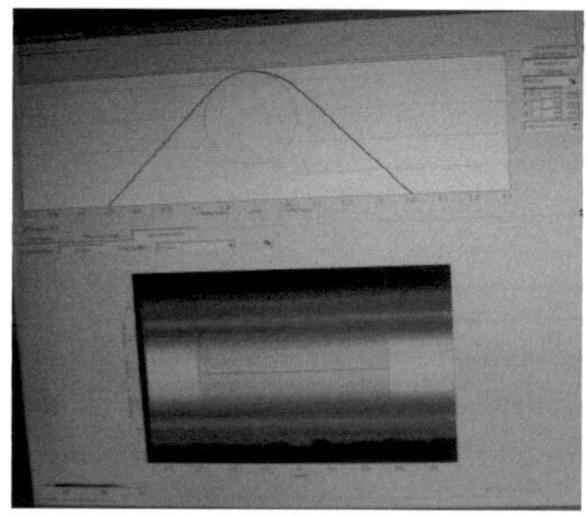

b) Data Acquisition and Processing

Figure 9. Measured Result of One Tooth of the Carbide Chaser After Preparation

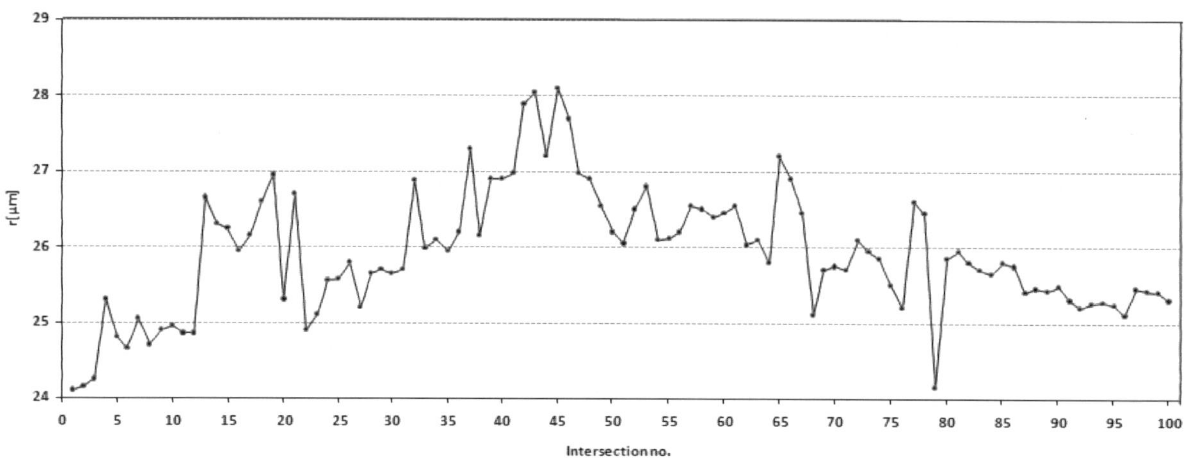

Figure 10. Rounding Radius of the Cutting Edge Along the Tooth After Preparation

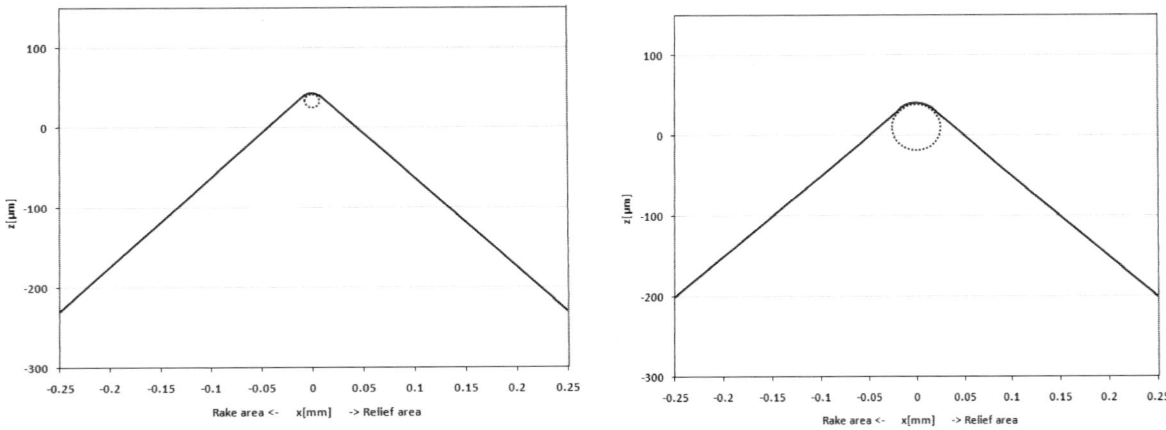

a) Before Preparation

b) After Preparation

Figure 11. Comparison of Radius of the Cutting Edge Before and After Preparation

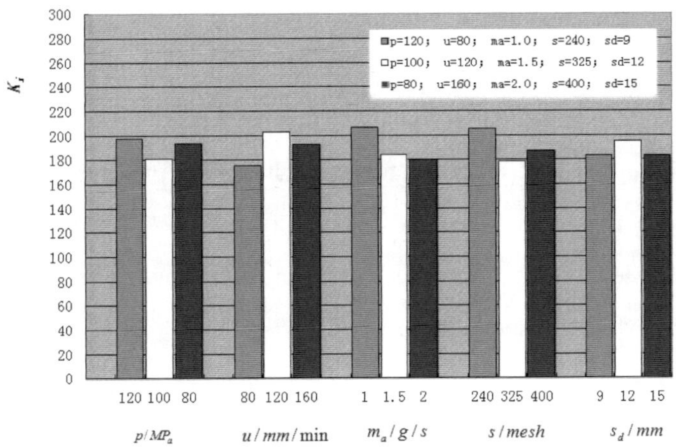

Figure 12. The influence of Process Parameter's Level on the Prepared Cutting Edge Radius

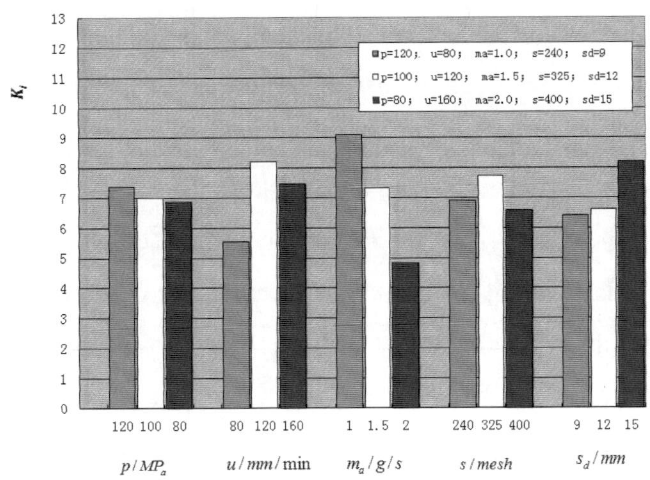

Figure 13. The influence of Process Parameter's Level on the Prepared Surface Roughness

2019 WJTA-IMCA Conference and Expo
November 11-13, 2019 • New Orleans, Louisiana

Paper

TURNING OF BRITTLE AND VERY HARD MATERIALS WITH ABRASIVE WATER JET

Fuat KARTAL

Mechanical Engineering Department, Kastamonu University, Turkey
fkartal@kastamonu.edu.tr

Abstract

In this study, turning materials which are very difficult to be turned by conventional methods due to their geometry and mechanical properties have been turned with abrasive water jet. Two different turning mechanisms have been developed for turning with abrasive water jet. Three-phases motorized turning apparatus for turning long work pieces at low and high turning speeds, and a stepper motor driven turning assembly for three-dimensional operations were developed. As a result of the study, HSS tool, glass goblet, drill bit, glass rod, polyethylene, glass bottle, bevel gear, screw shaft materials with abrasive water jet turning operations have been performed successfully.

Key words: Abrasive water jet, turning, complex and brittle materials

1. INTRODUCTION

Abrasive water jet (ASJ) treatment is classified as hydrodynamic treatment [1-2]. Abrasive water jet treatment, which is one of the unusual manufacturing methods, has advantages over other processing technologies [2-3]. With abrasive water jet, brittle, ductile, composite and low melting materials can be easily processed [1-4]. Due to the longer service life and the cutting speed, water jet grinding is less expensive than conventional technologies for processing such materials [1-6]. Although in steel and other common materials, the cutting speed of water jet grinding is actually lower than in conventional turning. This cutting technology is well suited for difficult-to-process materials: composites, glass, ceramics, titanium alloys, etc. Profile grinding wheels are also one of the possible applications with abrasive water jet rotation [1-7].

In this study, abrasive water jet turning of the samples which are difficult to turn by traditional method due to their geometry and mechanical properties was performed.

2. MATERIALS AND METHODS

2.1. Turning Experiment

The Schneider Altivar-12 220V-380V speed control board is used to control the speed in a 0.37 KW three-phase asynchronous electric motor. The special feature of this board is that it makes constant torque and speed of the motor by making adjustments such as rotation direction and speed with the three-phase motor. When machining with water jet, the turning direction of the lathe chuck continues to be constant according to the set speed value as well as there is no such factor as increase-decrease in the lath rotation speed. A special USB connection cable coded TCSMCNAM3M002P is used for programming the Schneider Altivar-12 speed controller card. The direction of rotation is determined by the direction button on the SoMove Lite v1.9.4.1 program user interface installed on the computer. In Figure 1, there is a special lathe mechanism which provides speed control with 3-phase motor.

Figure 1: Asynchronous Motor Turning Test Assembly Image

The most important feature of the lathe test apparatus is that the electric motor is higher than the level of the water jet water collection pond and that the splashed water particles cannot reach the electric motor. In the test apparatus, the use of a moving tailstock and a lathe chuck housing have given the apparatus a different feature.

2.2. Stepper Motor Turning Device

Another method for machining with ASJT is to turn the turning chuck to which the work piece is connected to produce three-dimensional parts at an angle of rotation. This method is the most convenient method to perform the same operation on different parts of the CNC program without removing the work piece from the machine and distorting the work piece coordinates (zero). With this method, products with complex geometries can be produced on more than one machining surface with different rotation angles. In this study, Nema 24 step motor with 1.8 degrees step angle with 3.1 NM torque is used. The 3-axis Toshiba TB6560 control and interface board, which is commonly used as a driver board, is used. Parallel Port Laptop provides Mach3 Mill software for programming the movement of stepper motor. The non-turning test setup using a stepper motor and four-legged 100 mm turning chuck is shown in Figure 2.

Figure 2: Step Motor Turning Test Assembly Image

2.3. Test Samples

The test specimens, the HSS turning tool in 10x10 mm glass cup, 14 mm diameter drill bit 30 mm diameter glass rod and polyethylene, glass bottles, screw made of case hardened steel shaft and bevel samples were selected (Figure 3). In particular, the selection of these materials is that they are both brittle and hard.

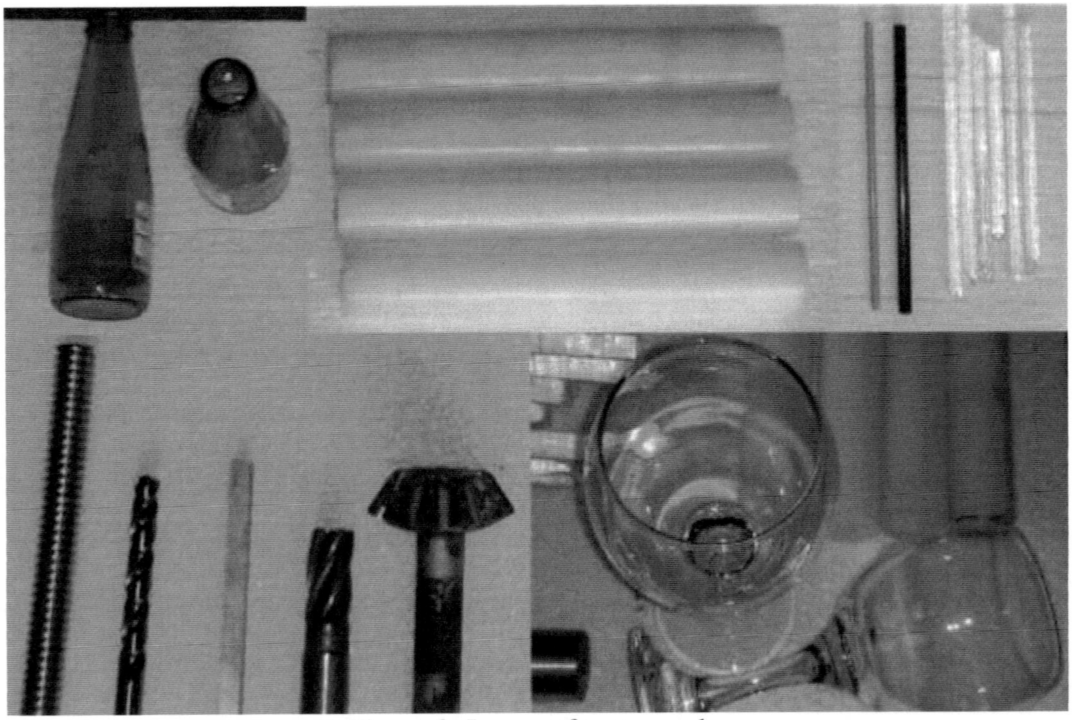

Figure 3: Image of test samples

2.4. Test Conditions

In this study, parametric study was not performed on parts machined with ASJT. The parameters were fixed based on experience (Nozzle feed 100 mm / min, abrasive flow rate 500 g / min, Nozzle distance 2 mm, pump pressure 3600 MPa, Nozzle diameter 1.3 mm, turning speed 20 rpm). In this study, the turning ability of hard, brittle and ductile materials, which are difficult to be processed by traditional turning method, was carried out.

3. RESULTS AND DISCUSSION

In this study, materials with different geometric shapes and mechanical properties are processed with asynchronous motor lathe or stepper motor lathe according to their properties with ASJT. Long and cylindrical work pieces that need to be centered with tailstock are machined using asynchronous motorized lathe. The work pieces, which are suitable for connection with short and four-legged lathe chuck, are machined with ASJT by using stepper lathe. It has been found that stepper lathe assembly is suitable especially for short materials. Square and ellipse materials are easily connected using a four-leg mirror. Both test devices were insulated against water. Figure 4 shows the moment of cylinder ring removal in the fragile glass goblet with ASJT. Figure 5 shows the ASJT-treated samples.

Figure 4: Image of Ring Extraction Process in Fragile Glass Goblet

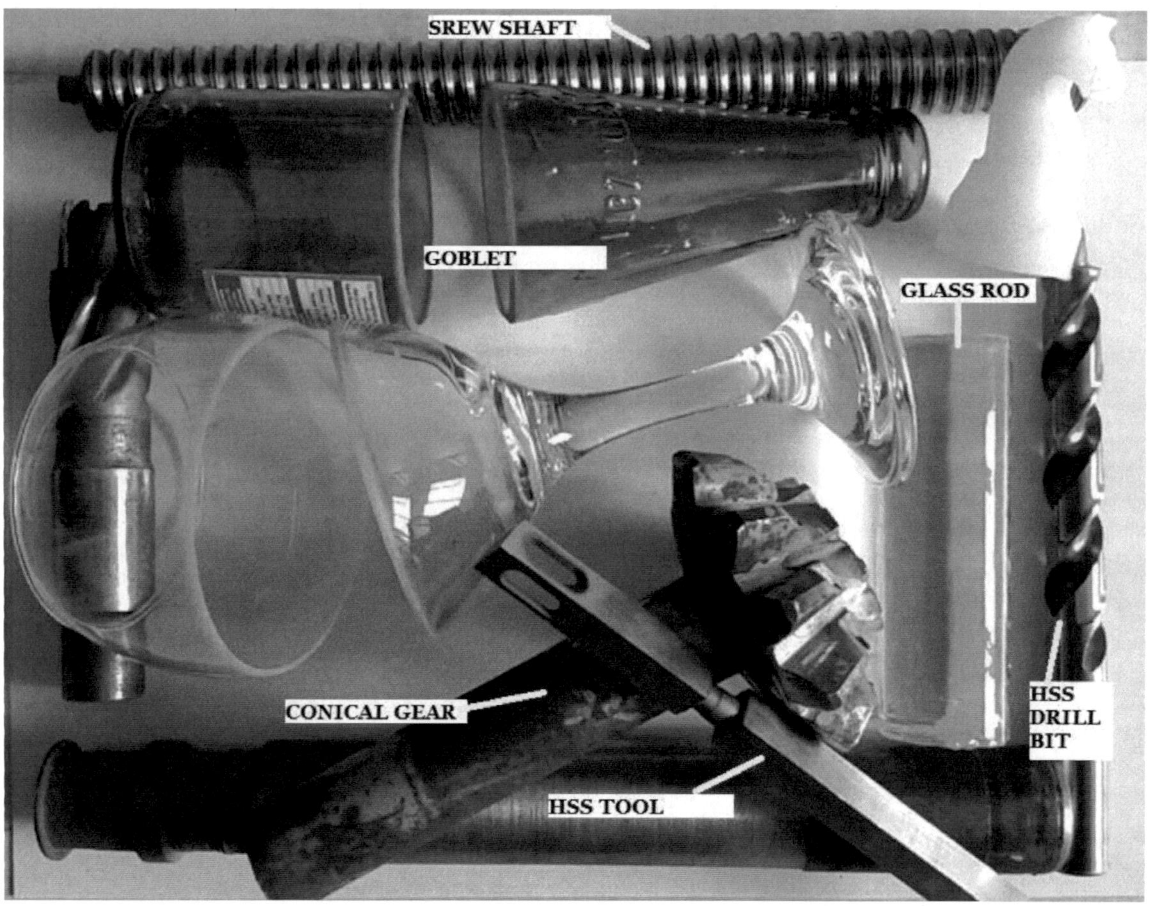

Figure 5: ASJT-Treated Test Sample Image

4. CONCLUSION

Within the scope of this study, work pieces selected from materials which can be difficult and even impossible to be processed by conventional manufacturing methods within the so-called limits have been successfully cut, diameter reduced, rough turning and forming processes with ASJT. In this study, no parametric study was performed. Findings obtained in this study;

- Different materials could be cut using the same nozzle.
- Square rectangles prism and ellipse geometry materials can be processed.
- The lathe chuck was rotated at 90 degrees to form a square, rectangular and prism sample surface without changing part coordinates.
- The ring was extracted from the very fragile glass.
- 14 mm diameter drill bit is reduced to 8 mm diameter.
- Bevel gear coarse turning process.
- Glass bottle cutting turning operation was performed.

REFERENCE

[1] Zhong, Z. W., & Han, Z. Z. (2002). Turning of glass with abrasive waterjet. Materials and Manufacturing Processes, 17(3), 339-349.

[2] Lv, Z., Huang, C., Zhu, H., Wang, J., Wang, Y., & Yao, P. (2015). A research on ultrasonic-assisted abrasive waterjet polishing of hard-brittle materials. The International Journal of Advanced Manufacturing Technology, 78(5-8), 1361-1369.

[3] Ramachandran, N., & Ramakrishnan, N. (1993). A review of abrasive jet machining. Journal of Materials Processing Technology, 39(1-2), 21-31.

[4] Choi, G. S., & Choi, G. H. (1997). Process analysis and monitoring in abrasive water jet machining of alumina ceramics. International Journal of Machine Tools and Manufacture, 37(3), 295-307.

[5] Korat, M. M., & Acharya, G. D. (2016). A review on current research and development in abrasive waterjet machining.

[6] Zhu, H. T., Huang, C. Z., Wang, J., Li, Q. L., & Che, C. L. (2009). Experimental study on abrasive waterjet polishing for hard–brittle materials. International journal of machine tools and manufacture, 49(7-8), 569-578.

[7] Hashish, M., & Craigen, S. (1990). U.S. Patent No. 4,955,164. Washington, DC: U.S. Patent and Trademark Office.

2019 WJTA Conference and Expo
November 11-13, 2019 • New Orleans, Louisiana

Paper

TECHNOLOGICAL ANALYSIS OF DIFFERENT ABRASIVE WATER JET MACHINING STRATEGIES FOR BLISK ROUGHING

T. Bergs, M. Schüler, K. Arntz, T. Herrig

Fraunhofer Institute for Production Technology IPT, Aachen, Germany

ABSTRACT

In recent years the production of BLISK (blade-integrated disks), to achieve an efficiency increase in civil aero engines, is highly demanded. To increase combustion efficiency, higher burning temperatures in combination with a higher over-all pressure ratio are necessary; the limiting factor is on the material side. From a production point of view, BLISK made of high-temperature resistant materials are hard to machine and therefore they are very cost and resource intensive parts. In order to remain competitive in the future and to combine higher productivity with better process reliability, promising manufacturing technologies like abrasive water jet (AWJ) machining need to be examined with regard to their technological and economic performance.

This paper reveals promising AWJ machining strategies for turbomachine production. Two machining methodologies have been introduced and discussed: cutting through the work piece and controlled-depth machining, often referred to as AWJ Milling. Technological analysis of experimental results is conducted in terms of necessary surface quality and simultaneously minimizing the machining time. Potential AWJ technology for BLISK roughing will be assessed and validated by machining a BLISK demonstrator made of Inconel 718 alloy. Finally, the experimental results will be evaluated and the feasibility of BLISK roughing by AWJ machining is discussed.

Organized and Sponsored by WJTA®

1 INTRODUCTION

The future demand for high-performance parts will increase significantly throughout all knowledge-based European industrial sectors like aerospace, automotive, energy, construction, consumer, and medicine. Integrated Turbomachine components like BLISK (= blade-integrated disk) are technologically characterized by a high material volume processed and invariable restrictions in quality requirements. An effective cost reduction of those parts concentrates on the machining process itself. The production time and machining costs must be reduced to remain competitive in high-wage countries. Consequently, the growing BLISK market calls for competitive production solutions, continuously improving existing processes and process chains.
Higher overall pressure ratio in combination with higher combustion temperatures are central strategies to increase engine efficiency. Modern airfoil blades are characterized not only by a complex geometry for the efficient aerodynamic implementation, but also by means of special high-performance materials to withstand operation loads. The next key improvement in engine technology is in particular the application of enhanced materials capable of amplified temperatures. Nevertheless, advances in material technology also require a further development in production technology to get the technological achievements of products economically and ecologically acceptable.

To dominate operational loads and increase thermal efficiency of turbomachines, high strength nickel-based alloys are in common use. Developments in metallurgy and manufacturing methods have led to one-piece integrated parts that improve mechanical properties and to reduce weight [1]. The BLISK design – successfully used firstly in military engines – is a promising alternative to fir-tree design. It found its way to latest civil aero engine concepts, where operational improvements must surpass costs in the end. These parts are typically milled from the solid; accordingly, a huge material volume must be removed in the primary roughing process step. The increasing demand of aero engines calls for higher batch sizes of these high-tech parts. Hence, the roughing process of BLISK made of nickel-based alloy Inconel 718 by conventional milling reaches its technological and economical limit. Turbomachinery components made of today's high-tech materials like the nickel-based alloy 718 are conventionally hard to machine and therefore require individual machining solutions. The machining approach is dependent of the available machining technology and the geometrical features of the work piece. Conventional machining methods such as milling uses several machining steps to obtain the final geometry and surface quality. For the rough machining process, a high material volume needs to be removed effectively from the solid before further finishing steps can be applied. The tool wear and therefore the machining costs are extensive for those rough machining approaches [2].

In contrast, abrasive water jet (AWJ) machining is already an alternative in the industry for cutting applications in a wide range of materials or compositions thereof. AWJ cutting is characterized by a high cutting performance. Especially when cutting high tech alloys or composite materials, process variants like AWJ milling promises to have huge overall advantages over other conventional and non-conventional processes [3]. However, the application of AWJ milling is not common in industry yet. During the last decades, water jet technology emerged applications, but it still lacks necessary developments supporting a wide industrial acceptance: Precise machining of complex geometries in thick work pieces like BLISK, needs commonly a huge expertise to handle the water jet characteristics [4]. This is mainly because insufficient software tools are used for complex tool path programming. Conversely, commercially available computer-aided manufacturing (CAM) software was originally developed for geometrically defined processes like milling and cannot be directly applied [5]. In general, the adaption of the machining strategy to the post process needs will further improve the precision and cutting performance of the involved processes and therefore judge between cost effectiveness and product quality. AWJ has shown to be a potential candidate to substitute conventional rough machining processes of BLISK [6]. Hence, the technological potential of different AWJ rough machining strategies urgently needs to be assessed.

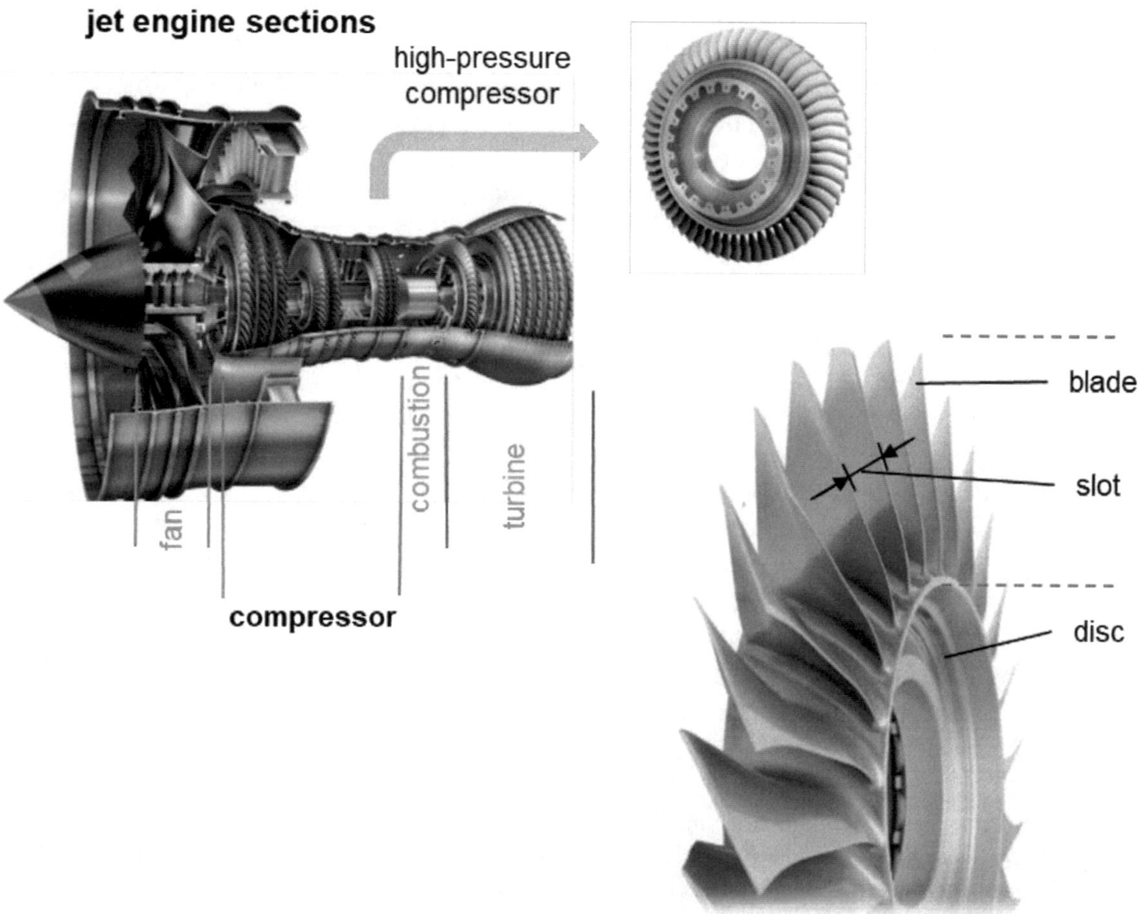

Figure 1: BLISK part within a modern jet engine compressor section
[Image sources: MTU Aero Engines AG, Sandvik AB]

2 MACHINING STRATEGIES FOR BLISK ROUGHING

The goal of rough machining is to approximately size and shape the work piece before it is refined in later processing steps. To economically optimize this process, a high material volume shall be removed quickly. In the case of BLISK manufacturing, it is of high importance to avoid contour damage since there are high-value integrated parts. Hence, precaution efforts, e.g. a defined safety offset to the final geometry, are needed for the intermediate machining steps. AWJ machining inhibits a large number of process parameters that encounters time-dependent non-geometrical uncertainties of the jet tool. Additional constraints must be considered to outbalance the process both technologically and economically. To achieve this goal, jet characteristics like the jet trail back has to be accounted for selecting the machining strategy. Industrially used AWJ cutting through (CT) is the simplest way of roughing a contour (**Figure 2**). The removed material is limited by the match of the straight-line cutout approximating the target geometry. However, as cutting depth increases or shape becomes more complex, additional constraints are given. Controlled depth machining (CDM), commonly referred to as AWJ milling, is theoretically the more flexible procedure to match a freeform contour but it lacks experience and its application is significantly more challenging [7].

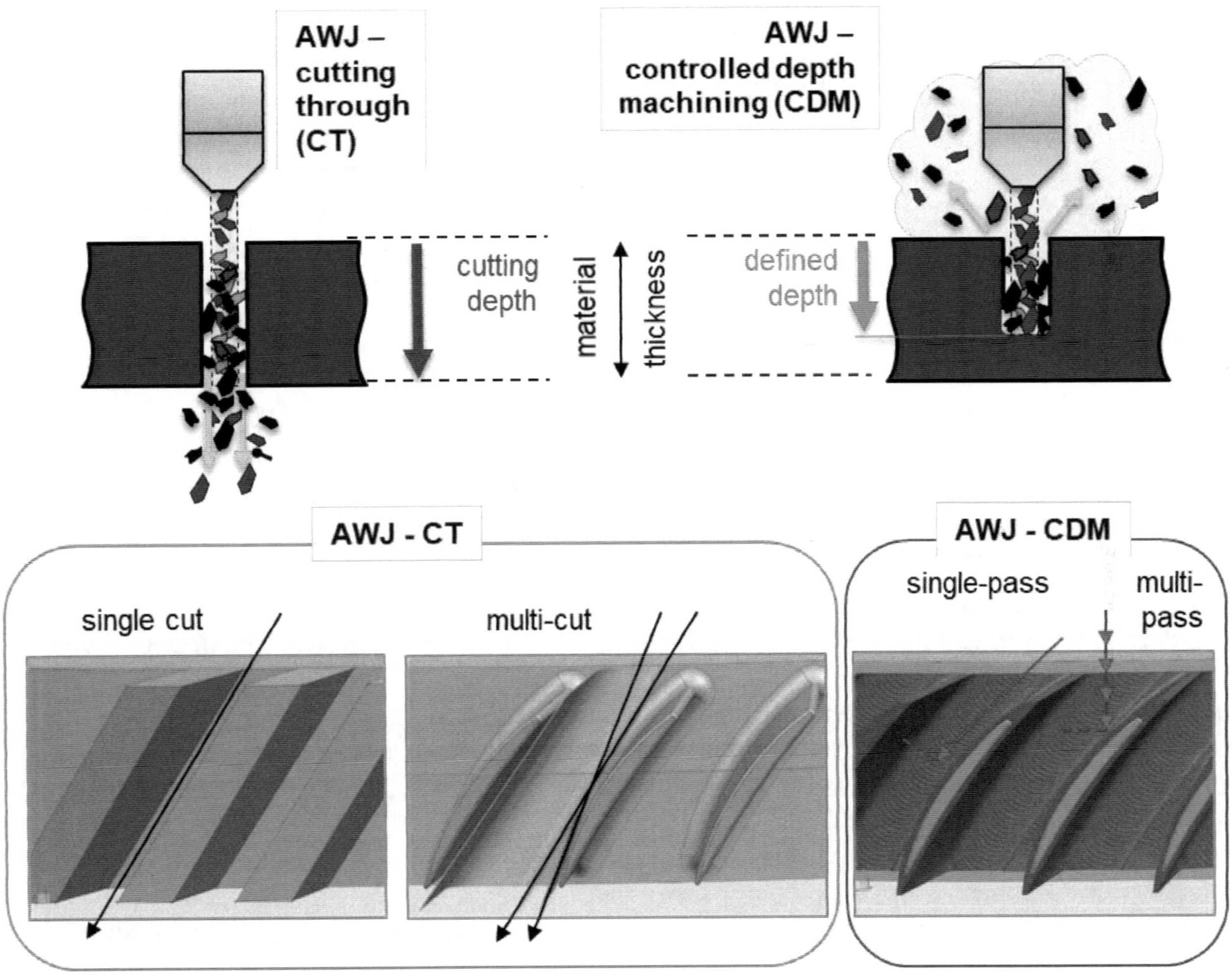

Figure 2: Potential of AWJ rough machining strategies

2.1 Cutting Through (CT)

Two cutting strategies can be distinguished: first, performing a single cut or second, performing several cuts to approximate the final shape. In both cases, the individual cuts go completely through the work piece. Cutting through (CT) is the most convenient AWJ machining application,

often used for 2D orthogonal cuts in industry. In contrast, the blade airfoil geometry demands a 3-dimensional approach by the application of a five-axis tool path. Generally, machining is limited to convex shapes since concave undercuts are not applicable for cutting completely through the work piece. Notice that the effective cutting depth is at least equal to the material thickness. The required cutting performance is extended further by tool paths angled to the work piece surface. The machining time will be affected negatively as it increases with increasing cutting depth. It can be expected that the jet trail back will increase, too. Especially at non-linear movements or going through edges of the tool path, the latter must be considered: Cutting a complex shape takes more time than performing a straight cut. Contour damage can be a result, especially on cumulative cutting depths; it goes along with the non-linear reduction of the cutting rate when cutting depth increases.
On the other hand, multi CT strategies make use of more than one single CT to approximate the final contour. As a positive result, the effective cutting depth will be a reduced amount of the previous single cut CT strategy. A decreased cutting depth offers several advantages such as less jet trail back and better surface quality. In other words, and for related instances, this enhancement to classic CT promises to be more effective in terms of material removal due to the applicability of higher feed rates in regions of effectively reduced cutting depth. Since the tool path becomes more complex, further constraints exist: The accessibility of the AWJ head must be checked as well as the clearance of the jet process itself not to affect neighborhood contours.

2.2 Controlled Depth Machining (CDM)

Controlled depth machining (CDM) shapes the work piece by superposition of numerous kerf profiles of defined depth. It is also referred to as AWJ milling due to its similarity to conventional milling processes. Here, two basic CDM techniques can be categorized: First, the machining of one single kerf, or second, the repeated application of several toolpaths of minor depth to achieve the aimed depth. For the first, almost the same constraints hold as for classic cutting, except that a well-defined depth must be machined. At given process parameters, the applied feed rate shall be constant. Due to reflections similar to jet trail back, the jet exits along the kerf leading to increased material removal [6]. Especially when machining complex contours, the AWJ must avoid contour damage by secondary material removal of the remaining AWJ. Applying an impingement angle in the feed direction could theoretically compensate the AWJ vector. However, the machine plant must be capable of control relevant process parameters. When using a 5-axis tool movement, even small impingement angles however lead to a decrease of the feed rate due to extended spatial travel of the machine axis system. CDM technique promises to make the AWJ machining process more efficient for 3D applications. Shape complexity can be extended over straight line cut outs, even convex shapes can be machined by lateral increment of the kerf profiles, e.g. when machining the airfoil upper part. If concave shape machining becomes necessary, access to the counterpart work piece side has to be granted, e.g. by turning the work piece. The transition to 3D CDM corresponding to conventional milling is implemented if several kerfs are concatenated to machine a certain contour. Instead of single kerf profiles to cut out a piece, the covered material volume is removed completely here line by line. Further, lateral and vertical tool paths are composed to machine a 3D shape. The time-dependent interaction of the AWJ tool with the work piece surface is similar to single-pass CDM cutting strategies. General for CDM techniques, both the contour complexity and the dynamics of the machine plant play an important role.

3 EXPERIMENTAL APPROACH

Several technological aspects limit the economic use of the AWJ manufacturing process within a process chain. Assuming presupposed standards for the interfaces, this technological analysis will focus on the technological aspects of AWJ rough machining. In particular of interest are the characteristics of the process parameters and their relationship to possible machining strategies. Different experimental plans for both machining strategies, i.e. cutting through (CT) and controlled depth machining (CDM), were carried out. Here, the objective concentrates on an economic roughing process to machine a contour close to the final geometry. Machining quality was analyzed by surface quality of the cutting edge lower area where the AWJ exits the work piece. In contrast to CT experiments, the CDM was analyzed by the geometrical characteristics of the kerf ground. ***Figure 3*** gives an overview of the individual measurement areas. In general, tow criteria are considered to evaluate the quality of the experimental results: first the achieved material removal rate (MRR) and second the cutting quality of the machined parts by means of surface roughness analysis. A statistic description of 2D surface measurements S uses quite more data even although conventional line measures such as roughness values R that were quite common for the analysis of periodic surfaces generated by milling. Contrary, the AWJ machined surface is more of a stochastic nature. Areal surface measures are definitely desirable to achieve robust results. Hence, the areal surface parameters S according to ISO 25178 were used for analysis of the surface quality [8].

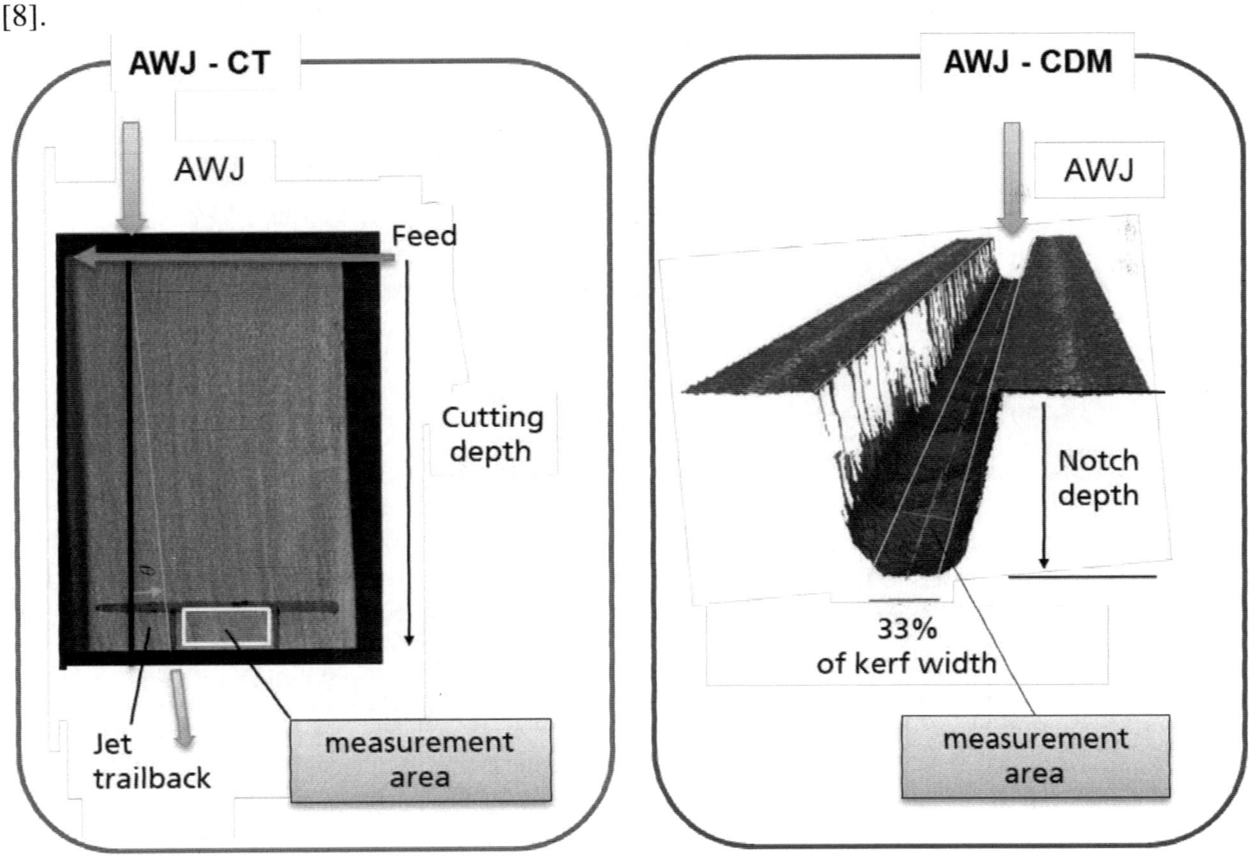

Figure 3: Analysis scheme of the test pieces

4 PROCESS DEVELOPMENT

An H.G.Ridder Type Waricut HWE P2030 5-axis AWJ machine tool was used for experiments, supplied with a UHDE Type 6045 intensifier pump. Parameter studies were conducted based on design of experiments methods and a cause-effect analysis of the machining strategies. As first experiment for cutting through (CT), a screening by comparative design was conducted to choose between two suitable AWJ configurations and to assess the machine tool for parameterization. Two standard AWJ set-ups of orifice diameter $\emptyset D_O$ and focus tube inner diameter $\emptyset D_F$ were used for the experiments for cutting different material thickness of nickel-based alloy 718:

- Configuration I: $\emptyset D_O = 0.25\ mm\ /\ \emptyset D_F = 0.78\ mm$
- Configuration II: $\emptyset D_O = 0.30\ mm\ /\ \emptyset D_F = 1.02\ mm$

The screening experiment revealed that AWJ configuration I of $\emptyset D_O = 0.25\ mm\ /\ \emptyset D_F = 0.78\ mm$ has showed the best overall results in both cutting performance and surface quality. Following, this configuration was used for optimizing the process parameters using a central composite design for CT. Furthermore, a study to assess controlled-depth machining (CDM) was conducted in analogy based on a full factorial design due to the significantly higher process complexity. The most important AWJ process parameters have been identified, namely the feed rate v_F, the pressure p and the abrasive mass flow $\dot{m}_A$. **Table 1** gives an overview of the AWJ parametrization range.

Table 1: *Optimization scope of AWJ parametrization*

AWJ machining	Pressure p in [MPa]	Abrasive mass flow $\dot{m}_A$ in [g/min]	Feed rate v_F in [mm/min]
CT: 20 mm	200 … 500	200 … 600	18.6 … 40.0
CT: 30 mm			11.7 … 24.6
CT: 40 mm			10.2 … 15.8
CDM	100 … 500	60 … 480	250 … 3000

The evaluation revealed that similar surface quality levels could be achieved by different AWJ process parameters. Consequently, the process parametrization was optimized for a defined cutting quality. This allows a better comparison of the results. The results revealed that the machined quality was in general contrary to the material removal rate (MRR). The only exception to this finding could be found for CDM with a feed rate beyond a critical threshold which yields a high surface quality even though increased AWJ cutting performance. This means in practice that if the feed rate is maintained beyond a critical threshold level, the MRR can be increased by an increase of the pressure and of the abrasive mass flow without drawbacks in the achieved surface quality.

Eventually, the findings were used to find a combination of AWJ process parameters that maximizes the MRR while optimizing the surface quality. The maximum surface height S_z and the root square height S_q were set to a defined target level. A quadratic model has been fitted for each response, which were combined to determine the composite desirability of the multi-response system. **Table 2** provides an overview how the proposed process parameters might be optimized for this purpose. Please note that the surface quality between CT and CDM differs due to the different machining principles. For CT, it can be seen that with increasing material thickness, a similar cutting performance can be achieved by increasing the abrasive mass flow within a constant pressure level while adapting the feed rate according to the material thickness. In contrast, CDM proposes both lower pressure and reduced abrasive mass flow values in combination with a significantly higher feed rate.

Table 2: *Experimental optimization of AWJ parametrization*

AWJ machining	**Pressure** p in [MPa]	**Abrasive mass flow** $\dot{m}_A$ in [g/min]	**Feed rate** v_F in [mm/min]	**Feed rate** v_F in [mm/min]	**Surface Height** S_q in [µm]	**Surface Height** S_z in [µm]
CT: 20 mm	387	397	26.65	MAX.	10	100
CT: 30 mm	390	435	17.46	MAX.	10	100
CT: 40 mm	397	600	10.44	MAX.	10	100
CDM	400	180	3000	MAX*.	15	150

5 CASE STUDY

A BLISK demonstrator geometry with typical geometrical features is used to debate the relevant rough machining specifications or roughing such a target geometry, a machining offset is needed to ensure that the target geometry is not damaged in any case. The computer aided design (CAD) data model of the BLISK was modified to generate the offset surfaces. The offset is approximates the original BLISK contour by an additional material volume added. For this investigation, the convex shape of the air foil blade is aligned by three main surfaces. In contrast, the backside of the blade is designed by a single surface. This modification is necessary to tolerate machining steps close to the final contour. All CAD modifications were done in Siemens NX software. An overall offset value of 0.6 mm was chosen, which is typical for conventional rough milling.

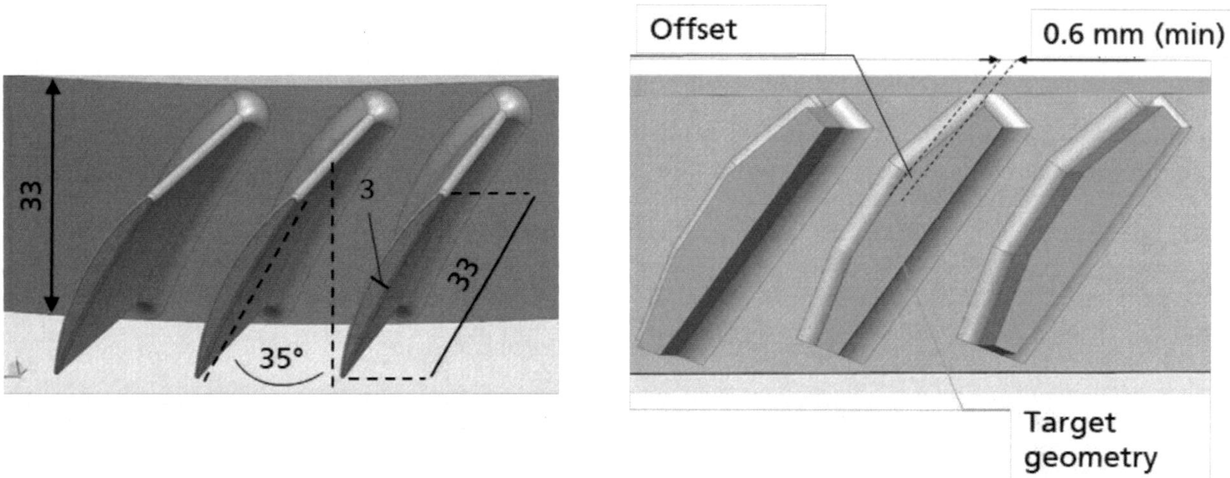

Figure 4: Geometry offset added to the contour of the BLISK CAD model

5.1 Proposed machining approach

In order to develop a proposal for optimized AWJ BLISK machining of given geometry, the developed tool path was applied to CT and to CDM so that in summary four different strategies were compared in terms of the removed material volume and the machining time. For CT, it is well known from the results of the parameter study which feed rate can be realized for a certain material thickness at a given quality. The time needed for single CT was measured as well as the total material volume to be removed by the corresponding cutout volume. It is highly important to sort the cut out pieces since the cutting quality decreases with increasing cutting depth. Therefore the remaining work piece exhibits the higher quality surface while the cut out material exhibits the inferior quality surface. Here, the 3rd cut represents the foremost cutout part. It is evident that for the second strategy, the multi-cut strategy, two cuts of defined depth have to be performed in the beginning (1st cut and 2nd cut). Because of the limited distance between the two airfoil blades, it was necessary to adapt the AWJ parameters so that the neighboring airfoil contour not to be damaged. In addition, the real machine system is limited in dynamics so that the high feed rate values could not be reached along the 5-axis tool path. The most promising way to machine a limited cutting depth is by increasing the applied feed rate. Note that the cutting contour of the voluminous single cut (3rd cut) was used in order to make the part fall out. Accessibility of the fourth cut was then possible so that the small portion in the lower part of the airfoil was removed as well. For the CDM strategies, the same procedure was followed but instead of cutting through, multiple kerfs of defined depth were machined until the comparable cutting depth was achieved. ***Figure 5*** gives an overview of the tool path design and the corresponding cut out pieces. The geometrical constraints of minimum cutting depth and allowed depth tolerance until neighborhood contour were estimated and rated according to the CAD geometry (**Table 3**). This helps to optimize the toolpath design and parametrization prior application. The CAM simulation is restricted to straight lined tool path and does not show varying cutting depth or interference.

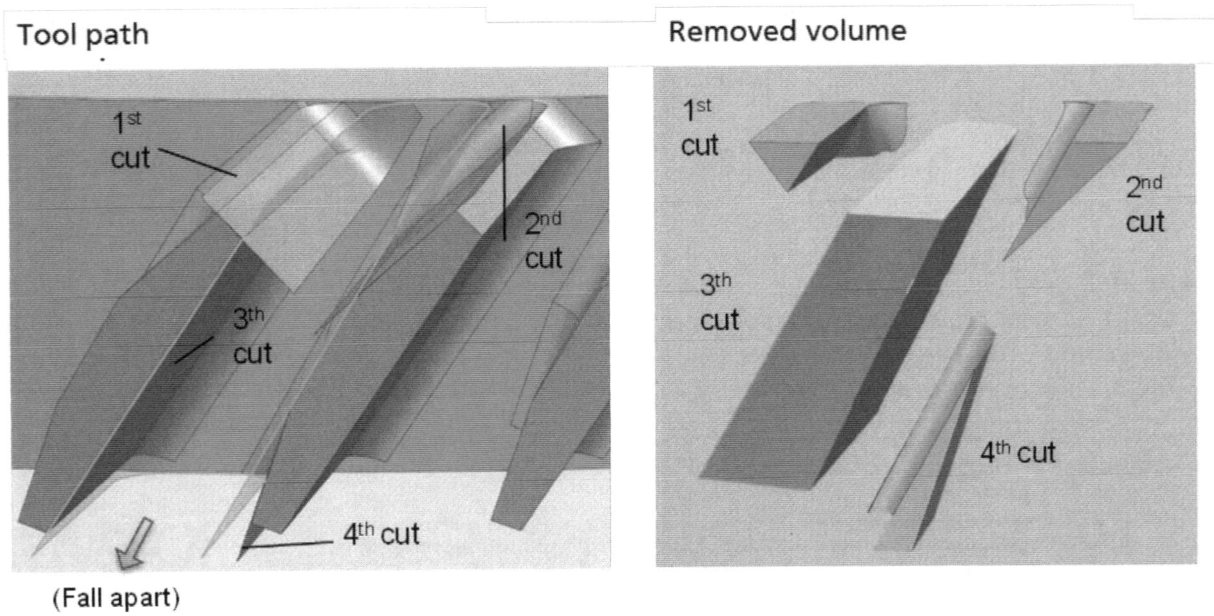

Figure 5: Tool path design and cut out pieces

Table 3: Geometrical features of the cut out pieces

Cut No.	**Material volume V** in [mm3]	**Tool path length l_T** in [mm]	**Minimum cutting depth t_{min}** in [mm]	**Allowed depth tolerance** in [mm]; **Relative to minimum cutting depth t_{min} in [%]**
1	458.53	44.09	5.45	9.77 (179%)
	855.79	40.94	15.14	19.26 (127%)
3	14 930.8	75.58	41.58	-
4	837.17	33.34	16.08	-

The categorization of discussed strategies was done by the stand-alone machining times that were calculated by means of the geometrical features and the proposed process parametrization. In **Table 4**, a comparison of the technological specifications is presented. Notice that the CT equivalent cutting times on the left side of were calculated by the experimental results based on the geometrical constraints of **Table 3**, e.g. the required CT time of cutting an equivalent material thickness. Even though the values in brackets are basically not applicable for CDM approaches, it helps to benchmark the different strategies. The equivalent MRR' of the CT cutouts was compared to the fixed MRR of the CDM kerfs assuming that the MRR of the individual notches remains constant even tough superposition. Here, 3D CDM means that the complete volume of the cut out pieces was machined completely whereas for 2D CDM a single cut comparable to CT was obtained by overlaying several notches up to a desired depth. Hence, a basic comparison of 2D CDM and CT was calculated by the kerf volume needed to be removed by obtained MRR. Accordingly, the result was then compared to the equivalent cutting times of CT. As the estimated kerf volume significantly influences the calculated 2D CDM machining times, a mean kerf width of 0.70 mm was used here that matches a typical CT kerf width. It was found for this comparison of stand-alone machining times, that choosing the machining strategy is strictly depending on the geometry to be removed. Hence, the lowest individual machining times marked in bold letters for each cut out geometry was selected. Because it is important that the 3rd cut will fall out of the work piece, CT was chosen instead by practical manner for this operation despite the slightly higher machining time. In contrast, 3D CDM takes much longer because of the huge material volume of the 3rd cut; CT is the proposed approach here. As a first comparison of machining this geometry, the additional machining time of the close-to-contour improvements (cut No. 1, 2 and 4) were approximately 25% higher than of the basic 3rd cut.

Table 4: Overview of the stand-alone machining times for the proposed strategies

	CT Equivalent		CDM			Strategy combination
Cut No.	**MRR'** in [mm³/min]	**Cutting Time** in [mm:ss]	**MRR** in [mm³/min]	**3D CDM Time** [mm:ss]	**2D CDM Time** [mm:ss]	**CT/CDM Time** [mm:ss]
1	(956.7)	(00:29)	1288	00:21	**00:16**	**00:16**
2	(1421.4)	(00:36)		**00:11**	00:41	**00:11**
3	3160.8	**04:43**		11:36	**03:29**	**04:43**
4	1129.9	00:44		00:39	**00:36**	**00:36**
Total time		06:33		13:16	05:02	06:15
Machining time of cut No. 1,2,4		01:49		01:40	02:52	01:32
Add. time in [%]		+ 27.8 %		+ 12.6 %	+ 30.8 %	+ 24.4 %

5.2 Machining of Demonstrator

Finally, a demonstrator geometry has been machined to validate the intermediate results of the findings under real circumstances.

Table 5 shows the machine parameters used according to the previous approach of using both CDM and CT strategies. The machining times for the individual cuts were recorded directly while machining. As practical result to the geometrical constraints, to the positioning movements of the AWJ head and to the NC-code optimization, the real machining times of the case study may differ from the previous findings of the theoretical approach (**Table 4**). Further, the parametrization of the AWJ process was optimized according to the AWJ machine constraints. In particular for CDM, the NC code was compressed to avoid unnecessary 5-axis movements of the machine in the region of turning points that helps to maintain the necessary constant feed rate. Prior machining, the federate along the NC tool path was simulated. This helps to precast the machining result leading to smaller the rejection rates of real manufacturing. Accordingly, the process parameters were tuned to maintain the necessary depth.

The machining process was optimized to avoid any damage of the target geometry. Notice that the minor differences in machining times was in both directions due to NC tool path optimization, e.g. by locate quality drawbacks of machining into the fall apart cut out pieces. For CT, there are two tactics to compensate the jet trail back. First, the AWJ impingement angle can be calculated to compensate jet trail back. This strategy maintains the same position level for both the entry and the exit side of the cutting line leading the AWJ curvature ahead of the feed direction. In contrast, a more easy way to program is to locally control the jet trail back by reducing the feed rate at the turning region. At a certain distance around the turning point, which should be at least the distance of the je trail back length, the feed rate is reduced. This allows the trail back to catch up before the AWJ changes its direction at the turning point. In analogy, the feed rate is increased when moving out the crucial region. Most machine controllers offer such a feature because the turning points of the tool path can be detected automatically and the feed rate is one of the easiest parameters to be controlled for cutting. Here, the feed rate at a certain distance of the turning point is reduced to 30% of the straight line feed rate value. In contrast, changing jet shape compensation by a tilted AWJ head is mostly not considered. For CDM, AWJ shape compensation may be indispensable due to the diverse material interaction and dynamic constraints. More important, the AWJ movements must maintain a constant feed rate to machine an equal depth. It must be ensured that the demanding high feed rates can be realized along the tool path; sharp turning points may be rounded to avoid feed rate drops along the tool path. The combination of the technological analysis with the individual machining times revealed that the combination of CDM and of CT strategies are a promising AWJ roughing approach. ***Figure 6*** shows the machined demonstrator geometry made of Inconel 718 alloy at real conditions. Here, the individual cut outs for the complete work piece were machined in a serial manner. In addition, by an extended machining time of cut No. 2, the unavoidable process irregularities of the bottom-end were relocated to the cutout piece. This helps to reduce individual machining times further. After the first two CDM operations were done, the 3rd cut out pieces were cut and falls apart allowing machine head clearance for the following 4th cuts.

Table ***5: Parametrization used for machining of demonstrator***

Cut No.	AWJ Strategy	Volume V in [mm^3]	Machining time T_m (m:ss)	Eff. MRR (mm^3/min)
1	CDM	458.5	0:31	≈ 886
2	CDM	856.8	1:05	≈ 790
3	CT	14930.8	4:25	≈ 3380
4	CT	837.2	0:32	≈ 1569
$\sum cut$		17083.3	6:33	≈ 2608

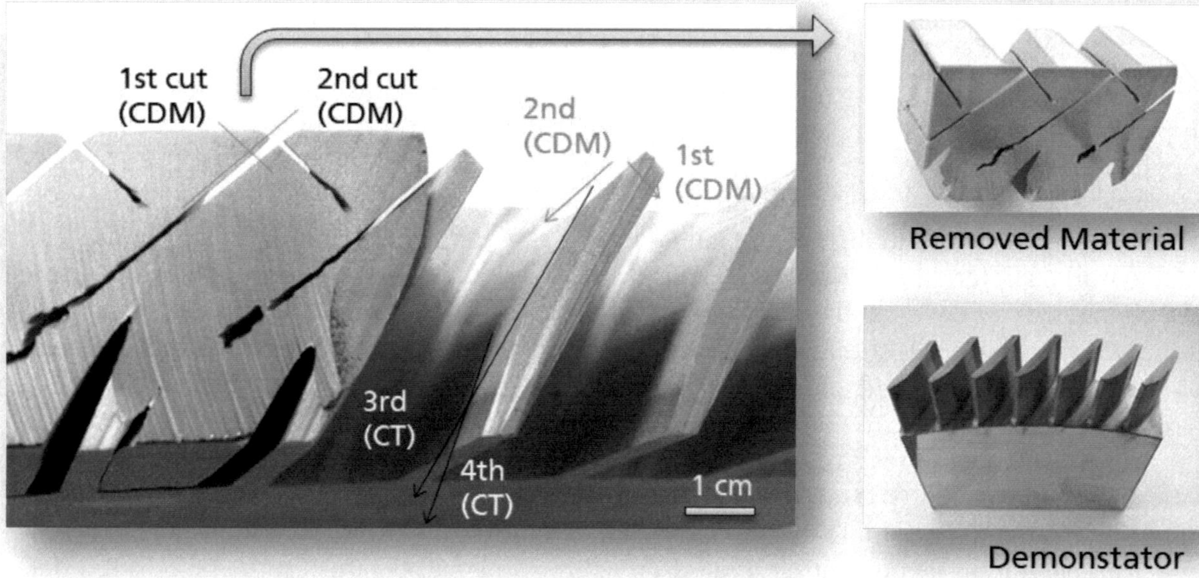

Figure 6: Machined demonstrator of applied strategies

6 SUMMARY AND OUTLOOK

In this paper, a proposal for optimized BLISK roughing by means of abrasive water jet (AWJ) machining was developed. The proposal is optimized in two ways: first, it is ensured that a satisfying surface quality is obtained and second, a strategy is identified which minimizes at the same time the production costs. In the following, two methodologies for AWJ BLISK roughing have been introduced and discussed: cutting through (CT) and controlled-depth machining (CDM). Parametrization studies have been set up for both CT and CDM by using design of experiment (DoE) methods. The use of areal surface analysis techniques have proven to be a good approach in terms of robustness of surface analysis, replacing conventional line measurements. In total, a huge optimization potential of AWJ process parameters for different demands has been identified in this work.

The comparison of different machining strategies confirms both the technological and economic advantages of AWJ machining for complex applications. It has been illustrated by experimental investigation that not only the right parametrization of the AWJ process influence the machining result. Performance, machining quality and costs must match both the AWJ machining strategy and the geometrical features of the work piece. The validation of this technological analysis of BLISK rough machining strategies has proven that CDM strategies facilitates close-to contour AWJ roughing. The practical machining of a demonstrator geometry confirms the validation of the parametrical findings.

In future work, optimization of the machining process should be considered. The technological findings should be extended by a cost analysis. There is probably a significant potential for improvement of the AWJ process by optimization of the abrasive-material interaction. Optimizing the impingement angle in an intelligent way could possibly minimize the amount of reflections yielding a more material removal efficient process. Future developments should target the development of appropriate CAM tools for AWJ machining in order to ease tool path planning and make AWJ an easy-to-use technology. In addition, a further optimization of the machine plants in terms of higher feed rates would be highly desirable. To summarize, the AWJ technology is a comparatively young technique that already produces high quality machining results as shown in this work but that still exhibits a huge potential for further optimization.

7 ACKNOWLEDGEMENTS

This paper is based on results acquired in collaborative R&D work, kindly supported by the ICTM International Center for Turbomachinery Manufacturing in Aachen.

8 REFERENCES

[1] Braeunling, W.J.G., 2009. Flugzeugtriebwerke, 3rd Edition. Springer Dordrecht Heidelberg London New York.

[2] Bußmann M., Bayer E., 2007. Market-oriented BLISK manufacturing - A challenge for production engineering, MTU Aero Engines, Munich.

[3] Hashish, M., 1987. Milling with abrasive waterjets: a preliminary investigation, in: Proceedings of 4th US Waterjet Conference, Berkeley, CA, pp. 1–10.

[4] Henning, A., 2007. Modellierung der Schnittgeometrie beim Schneiden mit dem Wasserabrasivstrahl, Dissertation, Universitaet Stuttgart, ISBN 978-3-939890-28-7, Jost Jetter Verlag, Heinsheim

[5] Klocke, F., Schreiner, T., Schueler, M., Zeis, M., 2018. Material Removal Simulation for Abrasive Water Jet Milling. Procedia CIRP 68, p. 541.

[6] Klocke, F., Schmitt, R., Zeis, M., Heidemanns, L. et al., 2015. Technological and Economical Assessment of Alternative Process Chains for BLISK Manufacture. Procedia CIRP 35, p. 67.

[7] Fowler G., 2003. Abrasive Water-jet - Controlled Depth Milling of Titanium Alloys, PHD Thesis. University of Nottingham.

[8] ISO 25178-2: Geometrical Product Specifications (GPS) - Surface Texture: Areal. ISO, 2012.

9 NOMENCLATURE

AWJ	Abrasive Water Jet
BLISK	Blade Integrated Disc
CT	Cutting Through
CDM	Controlled Depth Machining
MRR	Material Removal Rate
S_q	Surface Root Mean Square Height
S_z	Surface Maximum Height
$\dot{m}_A$	Abrasive Mass Flow
p	Water Pressure
v_p	Feed rate
CAD	Computer Aided Design
CAM	Computer Aided Manufacturing
V	Material Volume
l_T	Length of Tool path
t_{min}	Minimum cutting depth

2019 WJTA Conference and Expo
November 11 – 13 • New Orleans, Louisiana
Paper

IN VITRO RESPONSE OF OSTEOBLASTIC AND FIBROBLASTIC CELLS TO WATER-JET TREATED TITANIUM SURFACES.

A.J. Steeves, and F. Variola
University of Ottawa, Ottawa, Ontario, Canada

M. Vijay, A. Tieu, W. Yan, M. Xu and B. Daniels
VLN Advanced Technologies Inc., Ottawa, Ontario, Canada

ABSTRACT

Small-scale interactions between orthopaedic implants and biological systems impact how well devices will integrate in the human body. The response to these implant interfaces is fundamentally affected by the nature of the biomaterial that it is composed of, with key characteristics including surface chemistry and topography. It is generally accepted that rough implant surfaces have a more pronounced and beneficial influence on cellular activity than smooth ones. For this reason, methods to roughen metallic surfaces, such as sandblasting and acid etching, are currently used in the biomedical sector. In collaboration with VLN Advanced Technologies Inc, we propose to demonstrate the potential of Pulsed Water Jet (PWJ) as an effective method to modify the surface of titanium, the gold standard for biomedical implants, with a technique which only uses water. Inspired by previous collaborative work, we assessed the efficacy of PWJ modified titanium on the bioactivity of a fibroblastic cells (NIH-3T3), reflective of possible types encountered in current and future therapeutic implant implementations. Preliminary results have shown that PWJ modification of titanium is effective in the generation of surfaces with distinct nano- and micro-scale structural attributes which follow a gradient correlated with variable nozzle speeds. Preliminary results indicate cell-specific differential effects of these surfaces with larger fibroblastic NIH3T3 cells preferring balanced surfaces generated from a nozzle speed of 800 mm/s. Ultimately, the application of this environmentally-friendly and contaminant-free technique to industrial processes will permit to obtain implants that integrates in bone more rapidly while decreasing the manufacturing costs.

KEYWORDS
Titanium; Water jet; Surface characterization; Fibroblast; Osteoblast

1. INTRODUCTION

Whether due to age, lifestyle or by an accident, the replacement of teeth with dental implants has become a reliable and standard procedure with the number of prostheses designs extending beyond 1300 different implant systems.(Smeets et al.) Standard periodontal implants are composed of three main components: crown, abutment and implant (**Figure 1A,B**). Of critical importance is the implant ("screw") portion of the prostheses as it is the portion in direct contact with the jaw bone. The interaction between this screw and the bone ("osteo-") is the major determinant of the of implant success as this will influence how the implant will integrate/fuse with the patient tissue ("osseointegration").(von Wilmowsky et al.) In particular, numerous studies have demonstrated that the strength, and hence the performance, of osseointegrated implants is not only a function of the local bone quality and quantity (primary stability), but it also depends on whether fibrous encapsulation occurs (secondary stability).(Albrektsson et al.) Such phenomenon is produced when an interposing layer of connective tissue (e.g., collagen or fibroblastic matrix) forms between the surface of an implanted device and the surrounding bone, as a result of the wound healing process initiated by the introduction of a synthetic biomaterial. Such fibrous capsule is generally caused by micro-movements of primary unstable implants during early phases of healing, and is particularly detrimental for load-bearing devices such as the orthopedic (e.g. femoral stem) and dental (e.g. screw) devices, ultimately resulting in a compromised osseointegration.(Isaacson and Jeyapalina) In order to improve osseointegration of orthopedic and dental implants, significant efforts by the scientific community have focused on surface modification of implantable materials, in particular titanium, the gold standard in implantology.(Variola, Brunski, et al.) The goal is creating a surface capable of promoting the activity of bone cells while hampering that of fibroblasts (which represent the cells in oral gums), in order to secure strong secondary stability by eliciting bone formation and discouraging fibrosis encapsulation.(Variola, Yi, et al.) However, for dental screws, while the ability of limiting the formation of a layer of connective tissue is advantageous for the implant's osseointegration, it may also prevent, in turn, the creation of an effective biological seal at the implant-soft tissue interface. Such biological seal is considered a dominant factor from the clinical point of view to ensure the long success of peri-implant health.(Chai et al.) Peri-implantisis, a site-specific infectious disease, can in fact cause bone loss around an osseointegrated dental implant, thereby compromising its functions and performance.(Prathapachandran and Suresh) For this reason, achieving the capacity of modulating the response of fibroblasts by varying the surface properties of a titanium implant promises to unlock the key to engender dental implants that simultaneously discourage fibrous encapsulation but favour the creation of an effective interface with the gum. Previous work demonstrated that water jet treated surfaces have the ability to favour the proliferation of osteoblastic cells.(Steeves et al.) However, no studies to date have addressed the potential of water jet treated surfaces to modulate the response of fibroblasts for applications in dentistry.

The method and degree in which the dental implant component surfaces are treated prior to implantation has a critical effect on its success and longevity. As of 2015, the top methods of surface treatments currently employed sand blasting (~5%), SLA (~11%), acid etching (~15%) and spray plasma coating (~40%).(Jemat et al.) While these techniques have been successful in surface preparation, each comes with benefits and shortcomings such as production time, cost and the presence of contaminant by-product such as the embedment of abrasive particles from sand-blasting (**Figure 1C**). The use of a pulsed water jet (PWJ) may open the opportunity to not only fabricate thread geometries and surface prep in the same step, but also to do so in a sterile

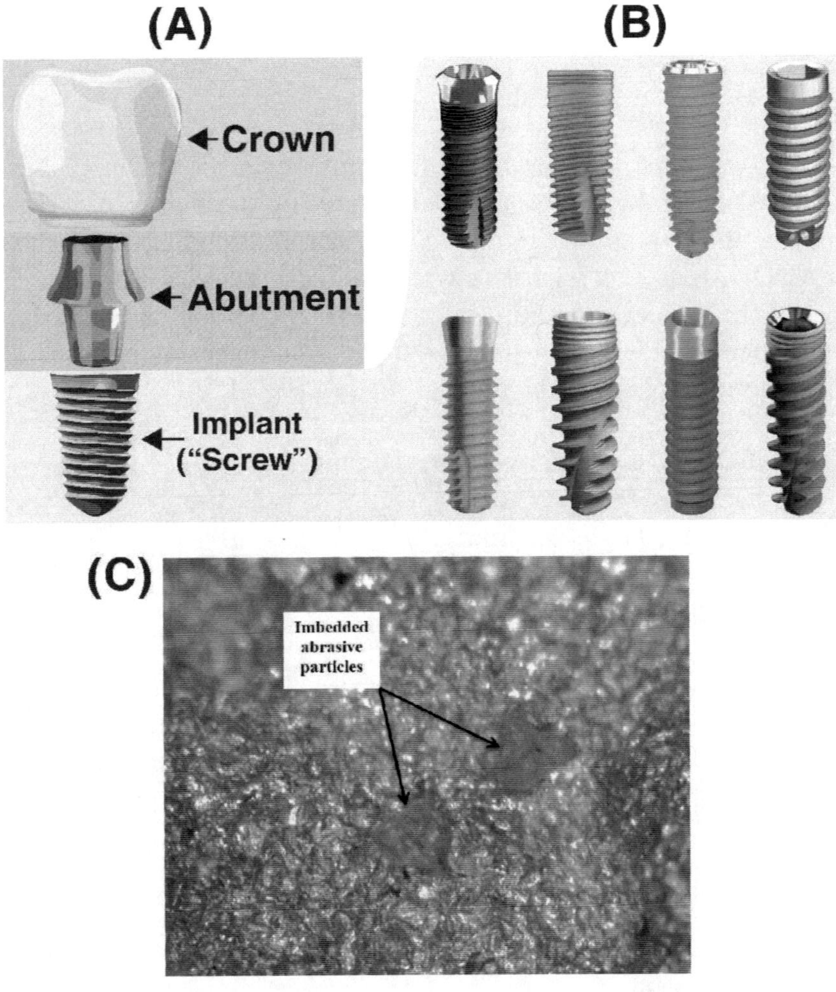

Figure 1. Periodontal implants. **(A)** Components of a common dental implant. **(B)** Examples of the various implant component designs. **(C)** Abrasive contamination from sand-blasted surface preparation.

environment (e.g., sterile water source, antibacterial additives) which would lead to a significant drop in manufacturing cost and possibility of compact fabrication machines in the dentist office. For these reasons, we decided to focus this work to generate an extended study to understand how the PWJ can generate optimal surfaces for the response of fibroblast cells (presented herein) with the additional response by osteoblastic cells and human stem cells currently in preparation.

2. EXPERIMENTAL SETUP AND PROCEDURE

2.1 Sample Preparation

30x20x1mm pieces of 99.2%-pure Titanium foil (Alfa Aesar, USA; Cat# 44484) were surface prepped in a PurePulse waterjet booth (Pratt & Whitney Automation, USA). The setup combined a heavy-duty turntable, high pressure pump (NLB Corporation, USA), robotic system with integrated HMI (FANUC, USA) as well as a water collection and filtration system (**Figure 2**). **Table 1** contains both the general operating parameters of the apparatus. **Table 2** contains the specific nozzle transverse speeds which generated the four experimental surface topographies ("E#"). Untreated titanium was kept as the control surface ("CTL").

Table 1. General operating parameters

Parameter	Value
Orifice Diameter (mm)	1.016
Pump Pressure (MPa)	69
FPWJ Frequency (kHz)	40

Table 2. Experimental nozzle transverse speeds

Sample Description *Experimental = "E"*	**Nozzle Speed** (mm/s)	**Stand-off Distance** (mm)
E45	450	3.5
E50	500	2.5
E60	600	2
E80	800	1.5

2.2 Surface Characterization

Visualization of the respective surfaces was achieved with the bright-field function of the alpha300 RSA system (WITec, Germany). Low magnification fields were visualized with an. To visualize focal adhesions, high-magnification 2x2 z-stack tiles were imaged with a 5x ApoPlan objective (NA=0.25, Zeiss, Germany). High magnification fields were visualized with an 20x Plan-Apo objective (NA=0.8, Zeiss). A series of images were taken at different positions in the z-axis ("z-stack") and were merged into a single plane with the depth of field "Stack Focuser" module of FIJI (ImageJ).(Schneider et al.; Schindelin et al.) Microscale roughness was assessed through the range analysis of the brightfield stacks.

Atomic force microscopy (AFM) analysis of surface topography was completed on the alpha300 RSA system (WITec). The Non-Contact (NC)-AFM mode, which resonates a nanometric tip

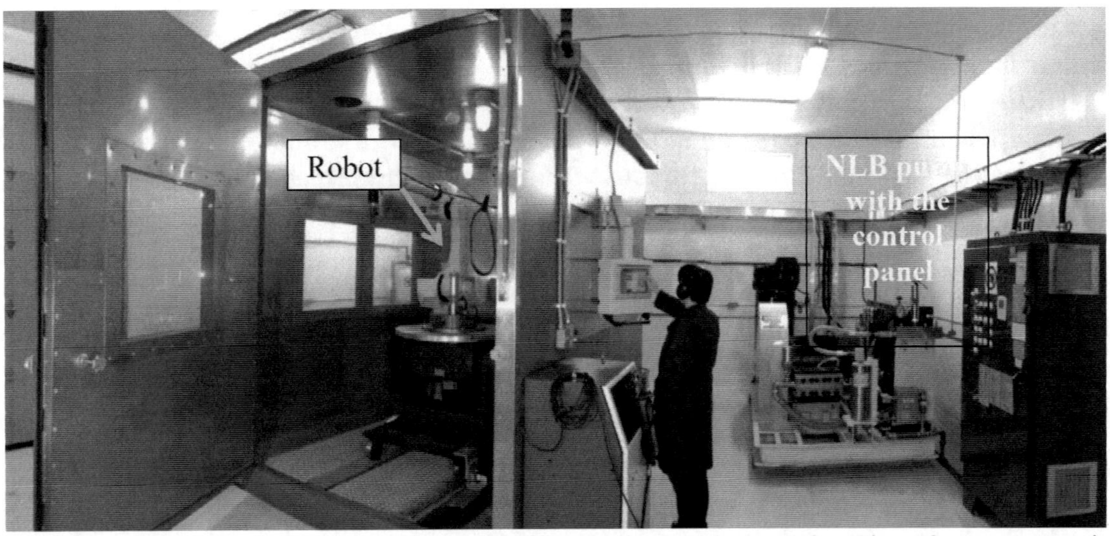

Figure 2. General view of PurePulse/PWJ experimental setup for Ti surface preparation

within angstroms above the surface as it raster scans, was used to prevent detrimental contact and wearing of the tip as a result of the significant differences in surface topography resulting from the surface erosion. 10 x 10 μm high-resolution scans (512 x 512) were imaged with the rectangular cantilever of the RFESPA-40 chip (Bruker, USA) with a tip that is characterized by a rotated pyramidal tip with a nominal radius of 8 nm, resonant frequency of 40 kHz and spring constant of 0.9 Nm^{-1}. Source files were imported into Gwyddion and processed for measures of surface area (SA) and roughness (RMS).(Nečas and Klapetek)

Raman and FTIR spectroscopy were carried out to give insight into characteristics of the titanium oxide layer (e.g., crystallinity). Raman spectroscopic analysis was carried out with the Raman module of the alpha300 RSA system. The pairing of a 50x Epiplan-Neofluar objective (NA=0.55, Zeiss) and Xtra 632 nm laser (35 mW, Toptica Photonics, Germany) was used for bulk surface spectra collection which consisted of 6 scans at 10 sec / scan. FTIR spectroscopic analysis was carried out with a Nexus 870 FTIR spectrometer (Thermo Nicolet, USA) equipped with the SAGA accessory (smart aperture grazing angle, 80° with respect to the surface normal; Thermo Nicolet) with an 8-mm-diameter opening. Spectroscopic information was collected in the 400–2000 cm^{-1} range with a 4 cm^{-1} resolution (256 scans per spectrum). Infrared data were analyzed by using the OriginPro software (OriginLabs, USA). Spectra were smoothed with the adjacent-averaging method and, after linear baseline subtraction, fitted with Gaussian functions to resolve secondary vibrational components according to previously published literature.(Variola, Nanci, et al.)

2.3 Cell Culture

NIH/3T3 (mouse fibroblast, ATCC, USA) cells were cultured in standard Dulbecco's Modified Eagle Medium (DMEM, Corning, USA) supplemented with 4.5 g/L glucose, L-glutamine, 10% fetal bovine serum (FBS, Gibco, USA), 20 U/ml^{-1} penicillin and streptomycin (Gibco). Culturing took place in a humidified 5% CO_2 water jacketed incubator at 37°C. Cells were detached from surfaces using appropriate volumes of 0.25% Trypsin (Thermo Fisher, USA) and incubated at 37°C for 5 min. Cells were collected and centrifuged at 190 g for 8 minutes. Successively, the supernatant was discarded, and pellet reconstituted in fresh culture media to desired concentrations specific to the experiment to a total volume of 500 μl per well.

After incubation at different experimental intervals, cells were fixed at room temperature with 4%-PFA for 10 minutes, washed with 1x-PBS and permeabilized with 0.25% Triton-X100 (Sigma-Aldrich) for 10 minutes. Nuclei were stained with NucBlue ReadyProbes (Thermo Fisher) and the actin cytoskeleton was stained via Rhodamine conjugated Phalloidin (Thermo Fisher). Multi-channel images were captured with an AxioObserver.Z1 inverted microscope (Zeiss) fitted with the Zeiss Filter Set 49 (Hoechst) and Chroma Filter Set 49005 (Rhodamine). Low-magnification tiles of the surfaces were imaged with a 10x A-Plan (Ph2) Objective (NA=0.25, Zeiss) for assessment of cell proliferation. Mid-magnification tiles of the surfaces were imaged with a 20x Plan-Apo (Ph2, DIC II) Objective (NA=0.8, Zeiss) for assessment of general morphology and cell spreading. Image collections were stitched together with AxioVision Mosaix Software (Zeiss) and focus stacked with Fiji. For the measurement of cell morphology, characteristics and proliferation, multiple custom pipelines were created in CellProfiler.(Carpenter et al.) Three main measurements were scrutinized for the cell structure, namely cell area, perimeter and form factor. Cell area was calculated by counting the scaled number of pixels within the identified cells. Perimeter was calculated by measuring the number of pixels around the boundary of the identified cell periphery. Form factor, calculated as $4*\pi*(\text{Area}/\text{Perimeter}^2)$, is the measure of a cell circularity: a value of 1 is representative of a perfectly circular cell while 0 indicates a linear cell. All data were aggregated, and statistical analysis performed in OriginPro.

3. EXPERIMENTAL RESULTS AND DISCUSSION

3.1 Surface Analyses

Surfaces were first visualized with upright light microscopy (LM) by 5x and 20x objectives in addition to gross morphology by handheld camera (**Figure 3**). Due to the thinness of the 20x objective focal plane versus the roughness of the materials, a sequential 3D stack of focal planes was obtained and an extended depth of focus (EDF) was generated with software to provide a 2D image with all portions in focus (**Figure 3C**). 3D data from these images were processed and microscale roughness was determined, during which it was determined that the E45 surfaces were critically rougher than all other surfaces, seemingly in a non-linear relationship to nozzle-speed (**Figure 4E**).

To better understand the impact of these morphologies at the scale that cellular structures would be identifying and interacting with (nanoscale; 10^{-9} m), we employed non-contact atomic force microscopy (NC-AFM) to for super-resolution visualization of surface topography. NC-AFM represents an in-direct method of cantilever-based nanoscale microscopy. The technique oscillates a microscale cantilever with a nanoscale tip tens of angstroms (Å) above the surface. Interaction with Van der Waals forces will lead to deviations in amplitude or frequency of the cantilever's oscillation which, as a non-destructive method, can be translated to topographical and phase data. 10 x 10 μm fields were imaged (**Figure 4A**) and processed for determination of nanoscale roughness, range and area enhancement resulting from treatment (**Figure 4B-D**). Nanoscale roughness was measured with the arithmetic mean (Sa) and surface area enhancement was estimated by the total surface area from the 3D data versus the projected area. For example, in a 10 x 10 μm field, the projected area is 100 μm^2 and a total surface area (due to structural morphology) of 115 μm^2 would result in a 15% enhancement. A completely flat surface would have an enhancement of 0%. While E45 presents with the greatest microscale range, E50 was found to have the greater nanoscale roughness and area enhancement (**Figure 4E**). This may be due to there being a balance between the generation of nanometric structures from the water slug spacing which resulted in elevated nanoscale roughness while preventing a drastic increase in deeper and irregular structures that would result in higher microscale roughness. As expected, E60 presented with higher nanoscale roughness and range, compared to slower speeds, however E80 saw a considerable loss. This may be due to the faster nozzle spacing the slugs far enough that the treatment creates a more uniform and with reduced incidence of deep erosion.

During the surface treatment and erosion of the material, the titanium oxide layer (TiO_2) is removed. While re-passivation occurs relatively quickly, the localized presence of heat effects from friction can lead to the generation of distinct oxide layer characteristics.(Vergara et al.) This TiO_2 layer is known to be critical in the cellular response, with specific mineral forms being able to influence behaviour.(Mazare et al.) To better understand the newly formed oxide layer, we used

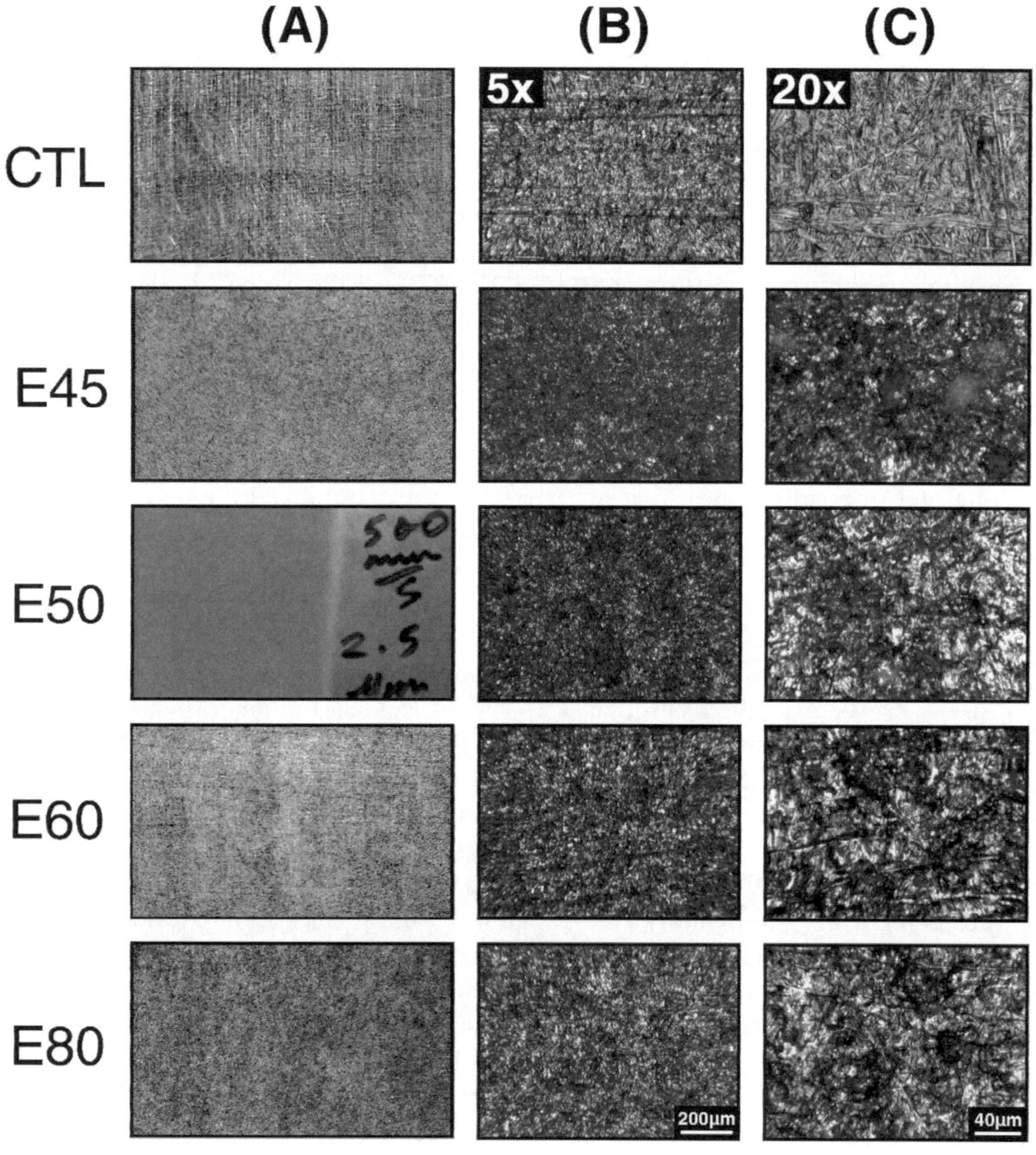

Figure 3. Visualization of treated titanium surfaces. **(A)** Gross visualization from handheld camera with enhanced contrast. **(B,C)** Light microscopy at (B) 5x and (C) 20x magnification.

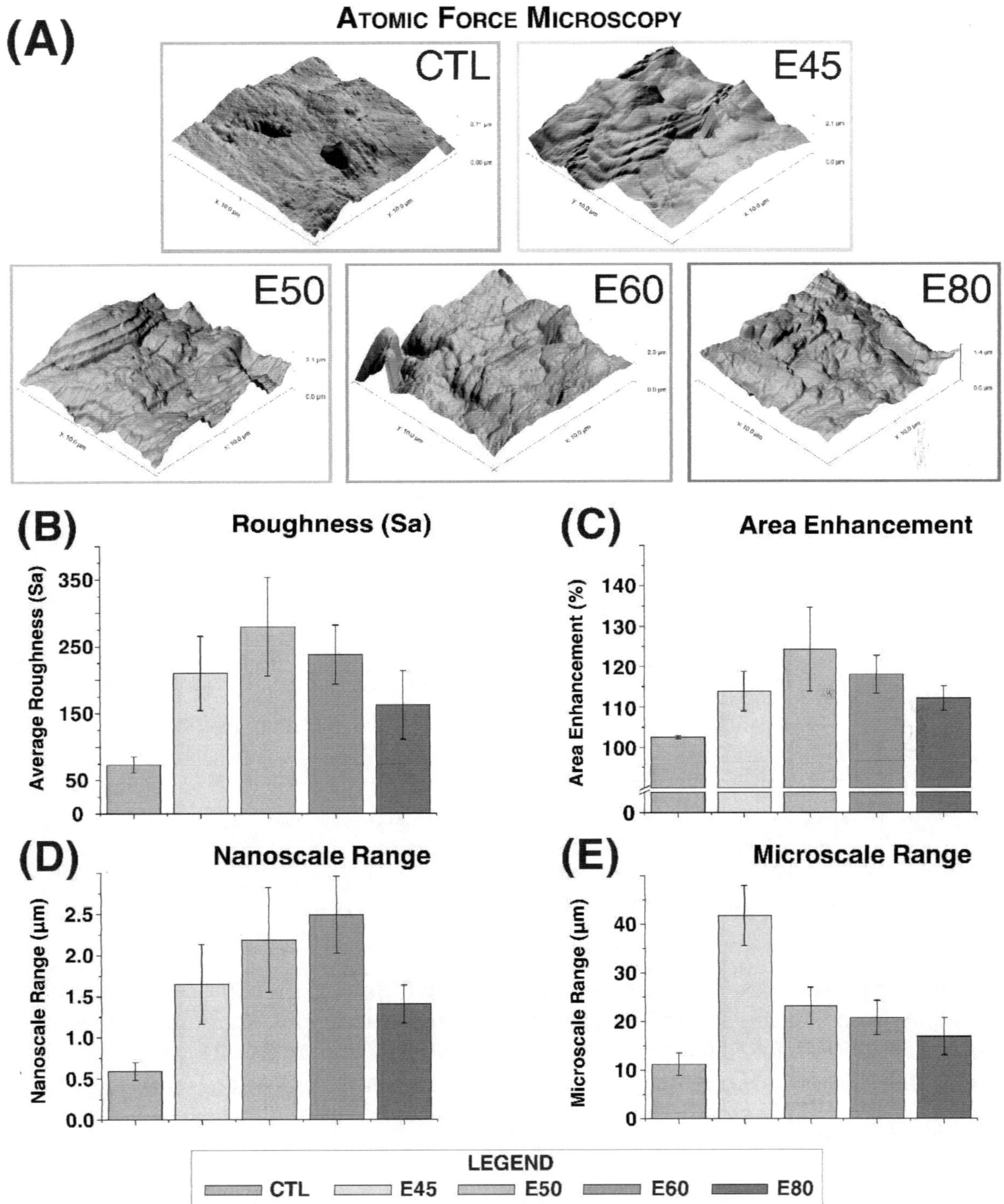

Figure 4. Analysis of treated surfaces with Atomic Force Microscopy. **(A)** Non-contact 3D-reconstructions of titanium surfaces. **(B-E)** Analyses of (B) roughness, (C) area enhancement, (D) nanoscale and (E) microscale ranges.

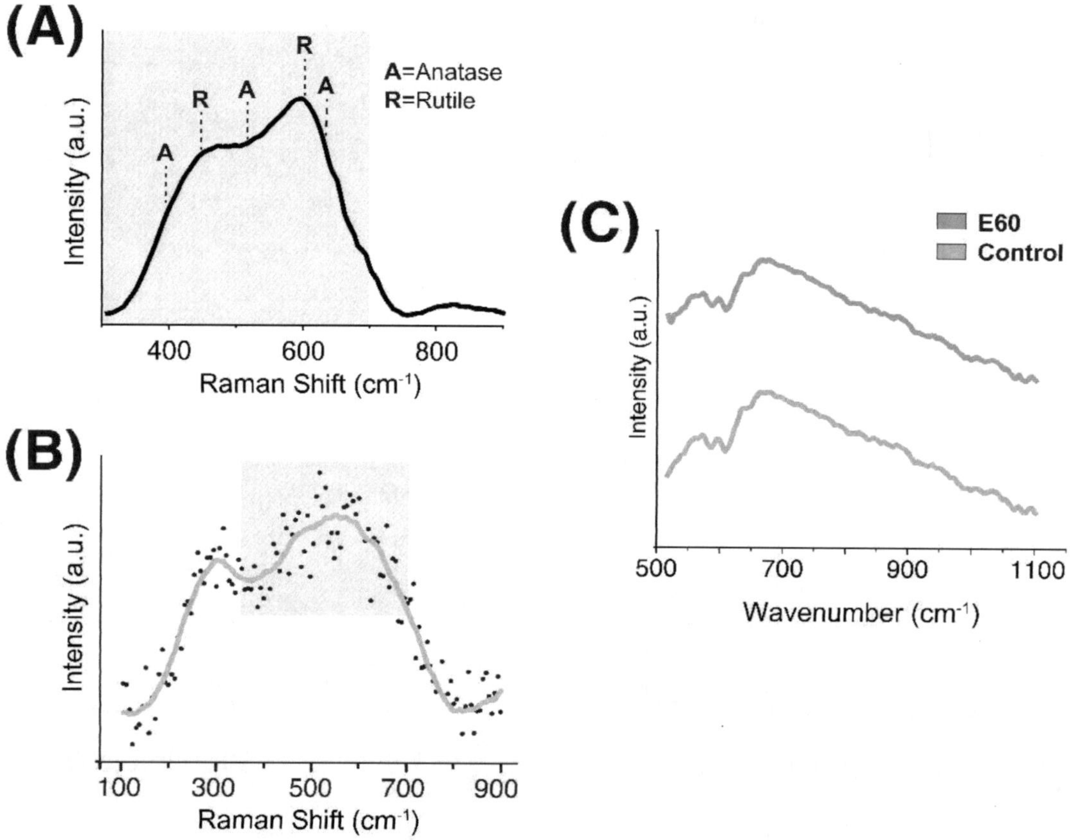

Figure 5. Spectroscopic analysis of titanium oxide films. **(A)** Example of an amorphous titanium oxide film containing Raman peaks seen in anatase and rutile forms. **(B)** Raman analysis of E50 with the blue region representing the field seen in (A). This represents an example of the waveforms seen on all surfaces. **(B)** Representative Fourier Transform InfraRed (FTIR) spectroscopic analysis of E60 and Control samples.

spectroscopic analysis through Raman and Fourier Transform Infrared (FTIR) spectroscopy to determine the crystallinity and thickness of the oxide layer. Raman analysis found that the fingerprint of all surfaces was found to be representative of an amorphous oxide layer, exemplified in **Figure 5B**, as seen with the presence of both anatase and rutile spectral bands (**Figure 5A**).(Hardcastle et al.) FTIR spectroscopy, which can be employed to determine the thickness of the oxide layer, showed no meaningful differences between control and treated surfaces.(Variola, Nanci, et al.) When taken together, spectroscopic analysis suggests that the use of PWJ does not impact the crystallinity of TiO_2.

3.2 Biological Analyses

Following the surface analyses, we concluded that the use of E50 surfaces were unnecessary and so only the remaining four surfaces were tested (CTL, E45, E60, E80). NIH-3T3 fibroblastic cells, representing those that exist in oral gums, were cultured on their respective surfaces for 48 hours to determine the short-term response to the titanium, mainly proliferation and cellular morphology. NIH-3T3 cells were found to have their proliferation perturbed on rougher surfaces (E45, E60) while the E80 surfaces showed comparable rates to that seen on control titanium (**Figure 6A-D,I**). This may be an indication that the cavities produced are incompatible with the size of a normally spread NIH-3T3 (CTRL = ~50-70μm) with those on the E45 and E60 surfaces presenting with sized of 25-40 and 35-50 μm, respectively (**Figure 6E-H**). The presence microscale roughness on E45 that is close to the minimum size of the cells is thought to lead to an inability to extend cellular structures and generate tension across the cytoskeleton for effective spreading. While the CTL surfaces were comparable to that found in E80, it is important to note that the initial surfaces which were composes of cold-rolled unpolished titanium. This CTL titanium presented with an heterogeneous morphology with irregular and stochastic damage from handling, generation and storage which in turn prevents any appropriate control of cell response. By treating these surfaces, at minimum, we are able to generate a comparably homogeneous surface with better predictability on the biological response which may then be translated to downstream dental implants.

4. CONCLUSIONS

In conclusion, our work indicates a cell-specific preference for titanium surfaces with surface morphologies generated with our pulsed water jet technology. While these surfaces consistently presented with amorphous oxide layers, the differences in roughness contributed to the preference of fibroblast cells (representing the gums of the mouth) for smoother surfaces (800 mm/s) with previous work indicating and osteoblast-like cells (representing bone of the tooth/jaw) prefer rougher surfaces (450 mm/s). These results point to the potential generation and surface preparation of periodontal implants in such a manner that it will promote the integration of said implants in bone and thus the success rate. Continuing work will investigate the effects that these surfaces may influence the behaviour of human stem cells, including the potential for their differentiation into new bone cells, which can assist implantation in individuals with health issues (e.g., osteoporosis) and advanced age.

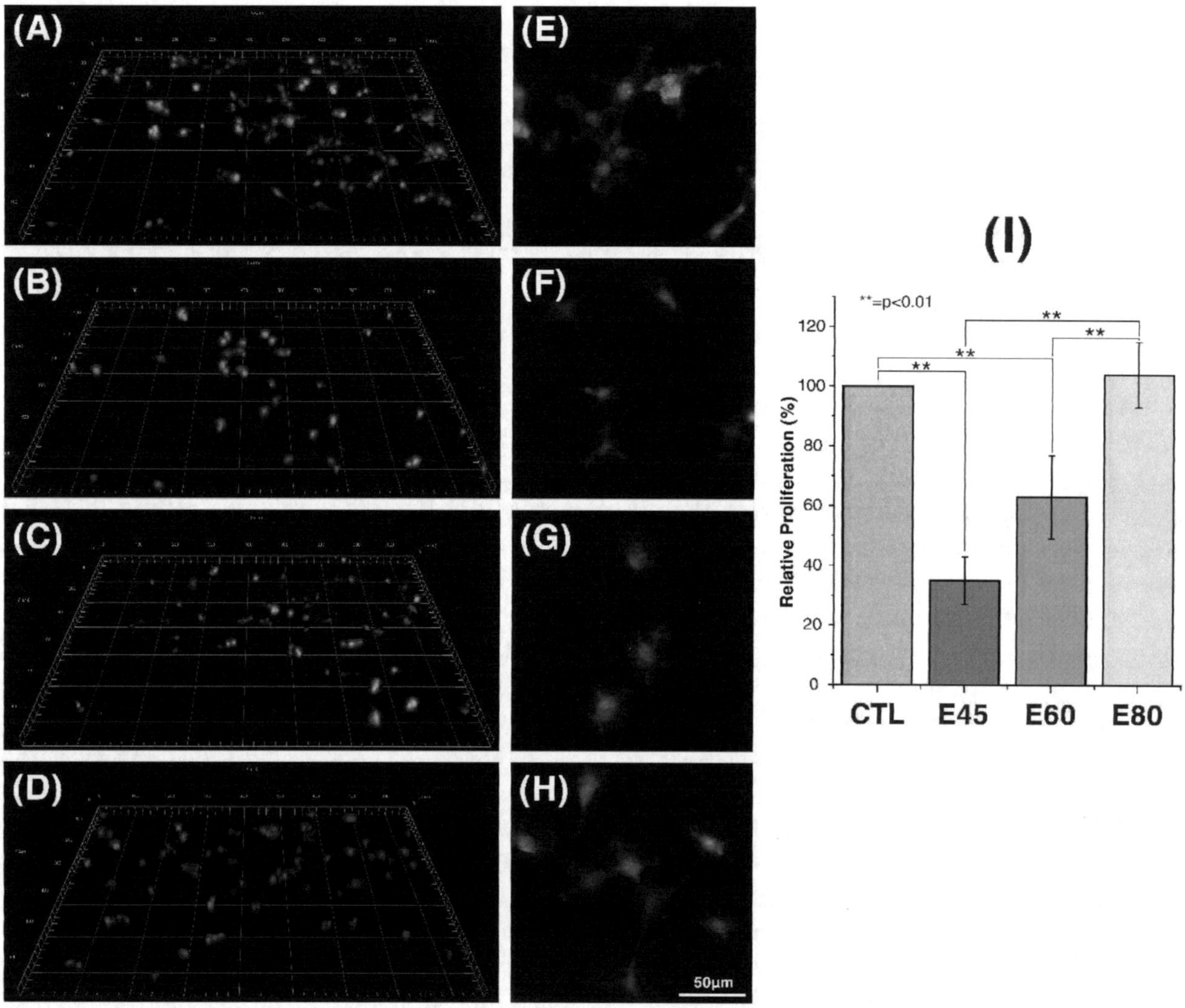

Figure 6. Cellular analyses of cells on treated titanium. **(A-H)** Fluorescence microscopy of NIH-3T3 cells at the (A-D) global level and (E-H) +50% zoom. Sample orientation is **(A,E)** CTL, **(B,F)** E45, **(C,G)** E60 and **(D,H)** E80. *Blue* represents the nuclei and *red* represents the actin cytoskeleton of the cell. (I) Proliferation analysis. One-way ANOVA; **=p<0.01.

5. ACKNOWLEDGEMENTS

This work was supported by the Ontario Centres of Excellence (OCE) Voucher for Innovation and Productivity (VIP1) grant, which strives to fund collaborative partnerships between company members in Ontario and post-secondary institutions. A.S. acknowledges the continued financial support from VLN Advanced Technologies and the Ontario Ministry of Training, Colleges and Universities through the Queen Elizabeth II Graduate Scholarship in Science and Technology (QEII-GSST).

6. REFERENCES

Albrektsson, T., et al. "Osseointegrated Dental Implants." *Dental Clinics of North America*, vol. 30, no. 1, Jan. 1986, pp. 151–74, http://www.ncbi.nlm.nih.gov/pubmed/3514290.

Carpenter, Anne E., et al. "CellProfiler: Image Analysis Software for Identifying and Quantifying Cell Phenotypes." *Genome Biology*, vol. 7, no. 10, 2006, p. R100, doi:10.1186/gb-2006-7-10-r100.

Chai, Wen L., et al. "The Biological Seal of the Implant-Soft Tissue Interface Evaluated in a Tissueengineered Oral Mucosal Model." *Journal of the Royal Society Interface*, vol. 9, no. 77, Royal Society, Dec. 2012, pp. 3528–38, doi:10.1098/rsif.2012.0507.

Hardcastle, Franklin D., et al. "Photoelectroactivity and Raman Spectroscopy of Anodized Titania (TiO 2) Photoactive Water-Splitting Catalysts as a Function of Oxygen-Annealing Temperature." *Journal of Materials Chemistry*, vol. 21, no. 17, May 2011, pp. 6337–45, doi:10.1039/c0jm03106b.

Isaacson, Brad M., and Sujee Jeyapalina. "Osseointegration: A Review of the Fundamentals for Assuring Cementless Skeletal Fixation." *Orthopedic Research and Reviews*, vol. 6, Dove Medical Press Ltd., 2014, pp. 55–65, doi:10.2147/ORR.S59274.

Jemat, A., et al. "Surface Modifications and Their Effects on Titanium Dental Implants." *BioMed Research International*, vol. 2015, Hindawi Publishing Corporation, 2015, doi:10.1155/2015/791725.

Mazare, A., et al. "Changing Bioperformance of TiO 2 Amorphous Nanotubes as an Effect of Inducing Crystallinity." *Bioelectrochemistry*, vol. 87, Oct. 2012, pp. 124–31, doi:10.1016/j.bioelechem.2012.01.002.

Nečas, David, and Petr Klapetek. "Gwyddion: An Open-Source Software for SPM Data Analysis." *Central European Journal of Physics*, vol. 10, no. 1, Feb. 2012, pp. 181–88, doi:10.2478/s11534-011-0096-2.

Prathapachandran, Jayachandran, and Neethu Suresh. "Management of Peri-Implantitis." *Dental Research Journal*, vol. 9, no. 5, Sept. 2012, pp. 516–21, http://www.ncbi.nlm.nih.gov/pubmed/23559913.

Schindelin, Johannes, et al. "Fiji: An Open-Source Platform for Biological-Image Analysis." *Nature Methods*, vol. 9, no. 7, July 2012, pp. 676–82, doi:10.1038/nmeth.2019.

Schneider, Caroline A., et al. "NIH Image to ImageJ: 25 Years of Image Analysis." *Nature Methods*, vol. 9, no. 7, July 2012, pp. 671–75, doi:10.1038/nmeth.2089.

Smeets, Ralf, et al. "Impact of Dental Implant Surface Modifications on Osseointegration." *BioMed Research International*, vol. 2016, Hindawi Limited, 2016, doi:10.1155/2016/6285620.

Steeves, A., et al. "IN VITRO ACTIVITY OF FORCED PULSED WATERJET (FPWJ) MODIFIED TITANIUM." *2017 WJTA-IMCA Conference & Expo*, 2017.

Variola, Fabio, Antonio Nanci, et al. "Assessment of the Titanium Dioxide Absorption Coefficient by Grazing-Angle Fourier Transform Infrared and Ellipsometric Measurements." *Applied Spectroscopy*, vol. 63, no. 10, 2009, pp. 1187–90, doi:10.1366/000370209789553129.

Variola, Fabio, John B. Brunski, et al. "Nanoscale Surface Modifications of Medically Relevant Metals: State-of-the Art and Perspectives." *Nanoscale*, vol. 3, no. 2, 2011, pp. 335–53, doi:10.1039/c0nr00485e.

Variola, Fabio, Ji Hyun Yi, et al. "Tailoring the Surface Properties of Ti6Al4V by Controlled Chemical Oxidation." *Biomaterials*, vol. 29, no. 10, 2008, pp. 1285–98, doi:10.1016/j.biomaterials.2007.11.040.

Vergara, L. I., et al. "The Role of Passivation in Titanium Oxidation: Thin Film and Temperature Effects." *Applied Surface Science*, vol. 187, no. 3–4, Elsevier, Feb. 2002, pp. 199–206, doi:10.1016/S0169-4332(01)00986-2.

von Wilmowsky, Cornelius, et al. "Implants in Bone: Part I. A Current Overview about Tissue Response, Surface Modifications and Future Perspectives." *Oral and Maxillofacial Surgery*, vol. 18, no. 3, 1 Sept. 2014, pp. 243–57, doi:10.1007/s10006-013-0398-1.

7. GLOSSARY

AFM: Atomic force microscopy. A method that employs a cantilever with a nanometric tip for in-direct or direct measuring of material properties (e.g., mechanical, topographical).
Crown: The portion of a dental implant that replaces the area of the tooth.
Implant: The portion of a dental implant that directly interfaces with bone and is often seen as variations of a screw.
Fibroblast: A type of cell that exists in connective tissue. Here-in as NIH-3T3s which represent the oral gum tissue.
Fibrosis: Thickening of connective tissues.
Fibrosis encapsulation: Compartmentalization and isolation of an implant, associated with implant rejection and failure.
HMI: Human-machine interface.
NA: Numerical aperture. An indicator of objective resolution.
Osseointegration: The process during which an implant fuses with bone tissue.
Osteoblast: Bone cell.
SLA: Sandblasted large grit and acid etched treatment.

2019 WJTA Conference and Expo

November 11-13 • New Orleans, Louisiana

Paper

FEASIBILITY OF SOFT TISSUE PREPARATION USING HIGH-PRESSURE WATER JET TECHNOLOGY

M. Mlinaric[1], M. Bauer[2], T. Hassel[1]

[1]Institut für Werkstoffkunde (Materials Science)

Leibniz Universität Hannover

[2]*retired from the institute*

A. Hocke[3], S. Hippenstiel[3]

[3]Department of Infectious Diseases and Pulmonary Medicine

Charité Universitätsmedizin Berlin

ABSTRACT

The preparation of living tissue from human bodies is of major importance for investigations on bodily functions in preclinical research. Most studies use animal tissue or immortalized human tissue. The results of these studies can only be limitedly applied to the human body. It is therefore of great interest for research to make living human tissue available for research. For the preparation of such tissue, specimens from resections have to be used. A central problem is the availability of resected tissue parts for biomedical basic and therapeutic research. The preparation of soft tissue is a great challenge due to its fragility and its unstable consistency. According to the current state of art, the first step of soft tissue preparation is a tissue-stabilizing instillation of resected parts. After the instillation, the tissue can be cut into slices by a vibratome. For a wide range of investigations, the tissue stabilization limits the possibilities of investigations, since cavities are instilled. To investigate the bronchial system for example, it is instilled with low-melting agarose, which has to be dissolved after the transection, which limits the investigation to the parts with large lumina. To overcome such disadvantages of soft tissue preparation, we have investigated an innovative method of cutting porcine lung tissue transections by means of waterjet cutting. It enables the atraumatic cutting of slices with a high degree of plane parallelism and a thickness of less than 500 μm.

1 INTRODUCTION

According to the current state of art, preclinical research on bodily functions in human or animal organisms is executed on extracted cells outside its natural tissue context or immortalized tissue parts. The transferability of the results to the living human organism is often low [1]. Examination methods for research on living tissue are very limited. Living human tissue is very rare but also living animal tissue has a limited availability for research [2]. The reason for that lies in the preparation of those specimens. To observe bodily mechanisms, optical microscopy techniques are used very often. To gain a microscopic view on the specimens and keep the tissue alive, it has to be prepared and cultivated in a defined manner. Although the preparation is necessary to investigate the tissue specimens, it limits the examination options technically. Thin tissue slices have to be cut for the cultivation. The available preparation techniques use vibrating blades to cut the resected specimens into these slices. To allow a clean cut and to get a smooth cut surface, the tissue is instilled with stabilizing media before cutting [3, 4].

In the context of this study, the feasibility of soft tissue transections by means of high-pressure water jet technology was investigated using porcine lung tissue. Due to its cavities and its spongy texture, lung tissue, like other soft tissue types, places high demands on the water jet process. For the mechanical preparation of lung tissue, low-melting agarose is used to fill the alveoles and to stabilize the tissue [3, 4]. The agarose fills the alveoles and limits the feasible analytical methods. The removal of the agarose after the transection is complex [5] and possible only in large-lumen parts, which limits the research to these areas. The concept of using a high-pressure water jet for the tissue transection is that the water jet applies low loads to the tissue, compared to mechanical cutting tools. Because of this low load, we should be able to dispense with a tissue stabilizing instillation.

As the institute of materials science has a wide range of experience in the field of water jet applications in biomedical engineering, we investigated the feasibility of soft tissue preparation with high-pressure water jet technology at our water jet laboratory. For example, we have formerly designed a water jet process as an osteotomy tool [6] and developed and manufactured myocardial stabilizing structures [7] and structures for the aorta replacement [8] using an abrasive water jet process.

To make resected tissue available for research by avoiding a tissue-stabilizing instillation, in cooperation with the Department of Infectious Diseases and Pulmonary Medicine, we developed a new method for high-precision tissue transections using high-pressure waterjet technology.

2 PROCESS REQUIREMENTS

The preparation of the soft tissue must be designed in such a way that microscopic examination can then be performed as best as possible. For this it is necessary that the samples have a homogeneous surface and a high degree of plane-parallelism. When cutting, the unsteady texture of the tissue is a major challenge. It is therefore necessary to gather a basic understanding of the behavior of soft tissue that is manufactured by a water jet. We decided to representatively use porcine lung tissue for our investigations. Porcine lung tissue has a high availability and has a spongy texture with a lot of cavities. For preclinical research on the immune system, the preparation of lung tissue is highly significant. Porcine lung tissue is very similar to the human lung, which is why our results can be transferred easily on the preparation of human lung tissue. It was to determine if the lung tissue can be cut reliably. To realize high-precision tissue cuts, three main points are decisive for the success:

- Homogeneous cut surface: To investigate the tissue specimens with a high depth of field and to make sure that none of the thin slices is teared apart during the process, it is of major importance to develop a process that cuts the tissue to plane-parallel slices with a homogeneous surface.

- Fixation of the soft tissue: To ensure a reliable cut, it is necessary to provide a safe fixation for the soft tissue. We investigated different strategies for the fixation.

- Low loads on the soft tissue: So that the two previous points can be accomplished, it is mandatory to keep the loads on the tissue that are inevitably applied to it by the water jet as low as possible.

To keep the loads low and to be able to cut precisely at the site to be examined, we decided to use an orifice with a small diameter. The productivity of the process was negligible, as the following investigations will take place over several hours and therefore take a long time in relation to the cutting process. For this reason, we chose an orifice with a diameter of 0.04 mm. The use of a small orifice ensures a narrow kerf and because of the low flow, only low loads are applied to the tissue, compared to an orifice with a bigger diameter. To make sure, that we can realize a precise cut over the complete cut surface, we measured the expansion of the water jet as a function of the distance to the orifice to determine the maximum specimen dimensions we can cut with a nearly coherent jet. As shown in Figure 1, the jet is highly coherent in a distance up to 30 mm from the orifice. We therefore concluded, that we will be able to precisely cut specimens up to a height of 30 mm.

Next to the use of an orifice with small diameter we tried to keep the jet's velocity low. The tissue is affected by venturi effect, resulting from the fast flowing fluid during the cut. The tissue tends to get sucked to the direction of flow into the kerf. This effect increases witch increasing flow velocity. The flow velocity depends on the water pressure, provided by the high-pressure pump. For that reason, we tried to keep the water pressure as low as possible while observing a satisfactory jet geometry.

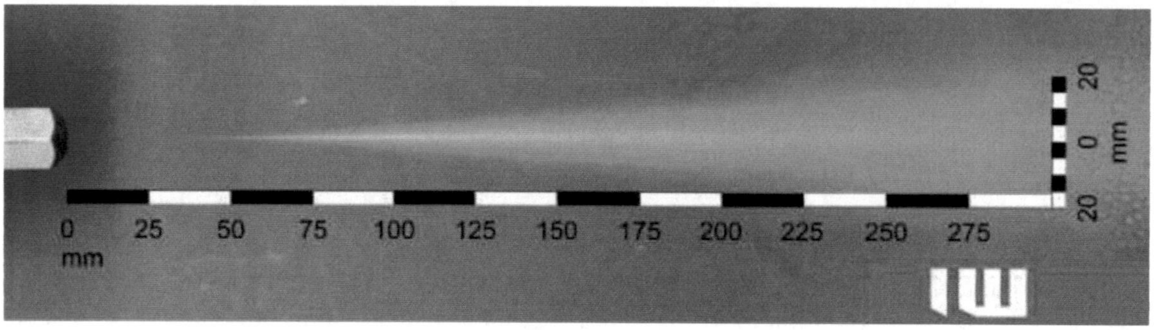

Figure 1. Water jet expansion at 70 MPa with a diamond orifice (d = 0.04 mm)

We repeated the measurements of the jet's geometry with pressures from 50 to 200 MPa by observing a maximum jet diameter above all pressures of 0.5 mm at a distance of 30 mm. This free jet's expansion is sufficiently low to ensure precise cuts. We identified the lowest expansion of the water jet around a pressure of 70 MPa, which is why we set a pressure range between 50 and 100 MPa for following test cuts.

3 TISSUE FIXATION

A safe fixation of the tissue is mandatory for the success of the cut itself. We tried to keep the fixation as easy as possible but the fixation of the specimen is a major challenge due to its spongy and soft texture. The lung tissue is very fragile because of its high amount of cavities surrounded by soft tissue. First experiments have shown that a mechanical fixation by piercing the lung provides a satisfactory result during the first cut. After dividing the lung in two pieces, no further fixation of the cut surface is possible, as shown in Figure 2. While cutting for the second time, the slice starts to wobble which has a negative effect on the plane-parallelism and the minimum thickness of the slice.

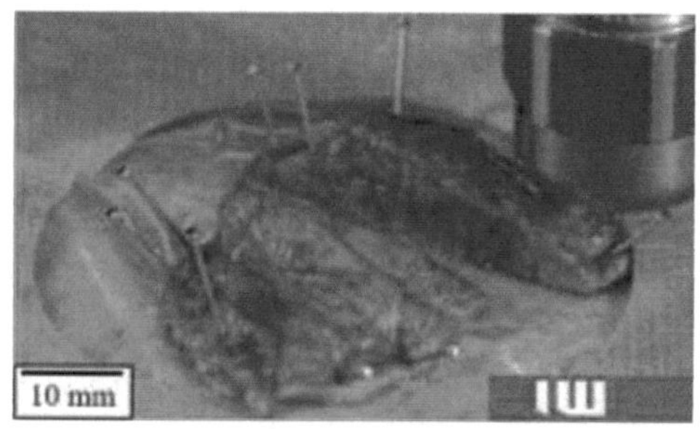

Figure 2. Test cut of porcine lung tissue using mechanical fixation.

To prevent the tissue from moving during the second cut we investigated two different fixation strategies. One uses low pressure over which the specimen is aspirated flat. The other approach uses a die cutter to pierce the lung and get cylindrical tissue specimens, which can be cut into slices afterwards. To cut slices from the cylindrical specimen, the tissue remains within the die cutter, which is fixated in a wedge-shaped clamping (see Figure 3).

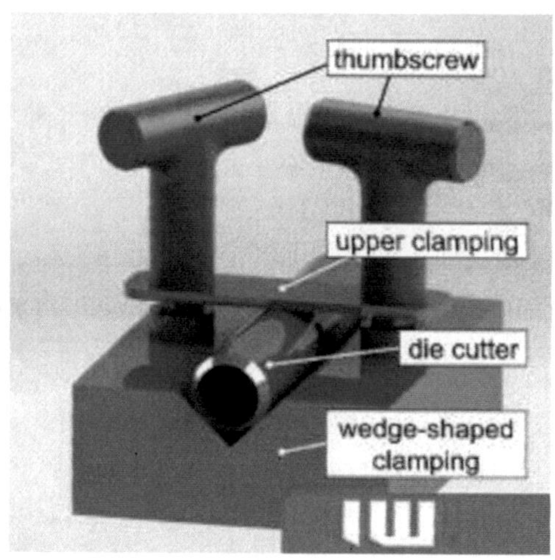

Figure 3. Mechanical fixation device with die cutter.

To investigate the mechanical fixation strategy using the die cutter, cylindrical specimens are die-cut from outer parts of the porcine lung. We used the outer parts, because there are relatively low cavities, which is why the tissue from the outer parts of the lung is slightly more stable. We placed the die cutter with the die cut tissue in the clamping and tried to cut thin slices from it. We first cut a slice to get a plane cylindrical end face. The second cut was to create a thin cylindrical slice.

It showed up relatively fast that the force that is applied to the tissue by the water jet causes poor cutting results. In addition to the impulse of the jet, the specimen is affected by the venturi effect, generated through the fast flow of the water jet. The cut slice gets sucked into the kerf and can generally not be recovered for further investigations. The few slices we recovered did not know a satisfactory cutting result. As seen in Figure 4, the die cut cylindrical tissue specimens already show an unstable geometry due to the soft texture of the lung tissue. Thus, as well as the mechanical fixation by piercing the tissue, the mechanical fixation by die cutting cylindrical tissue specimens from the lung is not suitable for the preparation of high-precision tissue slices.

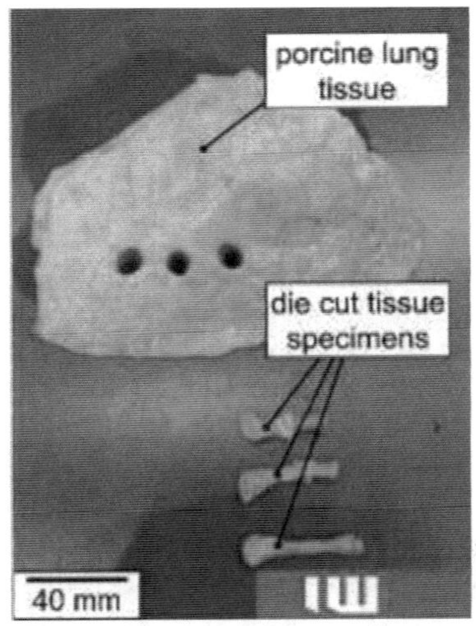

Figure 4. Die cut tissue specimens from porcine lung tissue.

Because a mechanical fixation does not lead to satisfactory cutting results, we came up with an alternative fixation strategy. We decided to build a fixation device that uses low-pressure to aspirate the tissue specimen from two sides (see Figure 5). The device is equipped with two suction grids. One is aligned horizontally and one vertically. It aspirates the specimen from the bottom and its side. The suction grid's hole pattern is quite coarse. We wanted to avoid the holes from plugging that there is no risk that the specimen unfastens during the cut.

Figure 5. Fixation of tissue specimens with low-pressure suction

For the generation of low-pressure, we used a VIPS 8 Type ejector of the company Landefeld GmbH. It uses the venturi effect, which is the reduction of fluid pressure when a nozzle increases the fluids velocity. The fluid flow is generated by pressurized air with pressures up to 0.7 MPa and an airflow up to 200 l/min.

For carrying out test cuts, we attached the specimen to the grids, through which the low-pressure is applied. Due to its non-plane-parallel structure, we aligned the specimen just roughly. The first cut serves to provide a plane-parallel surface on which the sample can be aligned more precisely. After the first cut, we removed the section and attached the plane cut surface to the vertical grid. That ensures the parallelism of the both cut surfaces. Due to the deforming of the tissue in order to the applied low-pressure, a deviation of plane-parallelism will occur.

We could affect the deviation by adjusting the low-pressure. It is to determine from which critical value a fixation of the sample is no longer reliable. During the first test cuts, we encountered two major problems with the cut slices: The tissue separates from the fixation device at high cutting pressures and the fixation causes nubs on the tissue surface.

The first problem addresses the suction of the water jet at high cutting pressures. As mentioned before, we observed a separation between tissue and its holder due to the suction of the water jet. Even with using the maximum air pressure for the ejector, this problem occurred from time to time at cutting pressures above 100 MPa. In order to that, we limited the cutting pressure to 100 MPa by keeping the ejection pressure at its maximum.

Figure 6. Nubs on the tissue surface caused by the suction through the perforated sheet

The second problem that showed up during the test cuts can be seen in Figure 6. The suction of the tissue holder causes nubs on the tissue surface, which remain after removing the tissue from its holder. These nubs emerge from the suction through the perforated vertical aligned sheet. They occur at the bottom side as well but are not mandatory because we want to gain a microscopical view of the cutting surface, which is aligned vertically. Despite the fact, the surface quality does not match the requirements yet, we could state a fundamental feasibility of the tissue fixation using low-pressure suction.

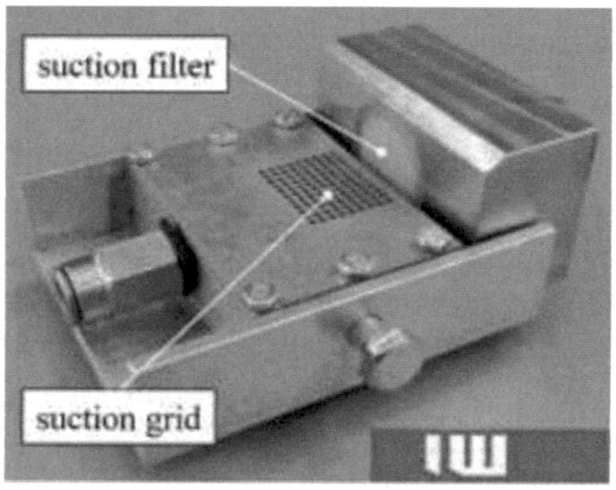

Figure 7. Modified fixation device with suction filter

To avoid the nubs on the tissue surface, we changed the perforated sheet for the vertical fixation with a commercially available suction filter of the company Fisher Scientific Inc. Figure 7 shows the modified version of the fixation device. The modification of the device by removing the perforated suction grid and installing a suction filter instead leads to a higher risk that the suction could plug. We therefore designed the holding that the suction filter is an easily replaceable part. To avoid clogging, it is replaced before every single cut.

The suction filter has a fine pored structure that leads to a uniform allocation of the suction force at the tissue surface. With the usage of the suction filter, we observed no more nubs. Next to avoid the suction filter to clog, a replacement of the suction filter against with a filter of another mesh can be useful to fix other tissue like liver or spleen.

4 CUTTING RESULTS

To evaluate the cutting result, we carried out further test cuts with porcine lung tissue. After cutting tissue slices, we used a confocal laser scanning microscope to get three-dimensional geometric data of the slices and to contactless measure thickness and plane-parallelism. Initially, we did not focus the minimum slice thickness but the homogeneity of the cut surfaces. After designing the process to achieve homogeneous cut surfaces, we minimized the slice thickness.

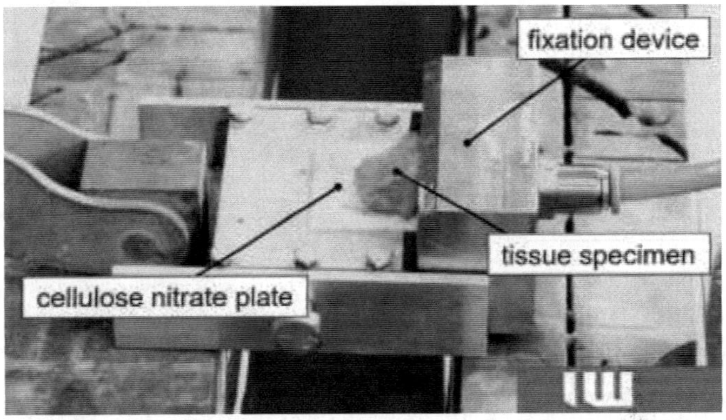

Figure 8. Porcine lung tissue attached to the fixation device with cellulose nitrate plate.

Resulting from the measurement of the jet's expansion, we varied the cutting pressure between 50 and 100 MPa. Compared to common water jet applications, this pressure range is quite low, which is advantageous for the application of soft tissue cutting. As described before, the low cutting pressure leads to a low flow velocity. Because of the low flow velocity, the suction from the water jet is low as well and we did not stand the risk of a separation of the tissue from the fixation device. As we presumed, a pressure of 70 MPa leaded to the most homogeneous cut surface. Although we kept the cutting pressure to a minimum, we mentioned that the water jet causes a vibration of the tissue, which leads to an undulated cut surface. To damp the vibration we additionally placed a plate made of cellulose nitrate under the tissue specimen, which successfully decreased the vibrations (see Figure 8). All further cuts were performed with a cellulose nitrate plate under the tissue specimen.

In the next step, we investigated the optimum feed speed while cutting with a pressure of 70 MPa. We varied the feed speed between 0.17 and 5 mm/s. To evaluate the cutting result, we determined the tissue's condition microscopically. The investigations showed up, that feed speeds below 2.5 mm/s lead to an atraumatic cutting surface. The higher the feed speed, the more damage could be observed at the cutting surface. Cutting extremely slow leads to disadvantageous surface qualities as well, as the tissue influences the water jet more often due to its small statistical movement. The optimum feed speed was between 0.42 and 1.17 mm/s. Figure 9 shows a cut slice and a picture taken with the laser-scanning microscope.

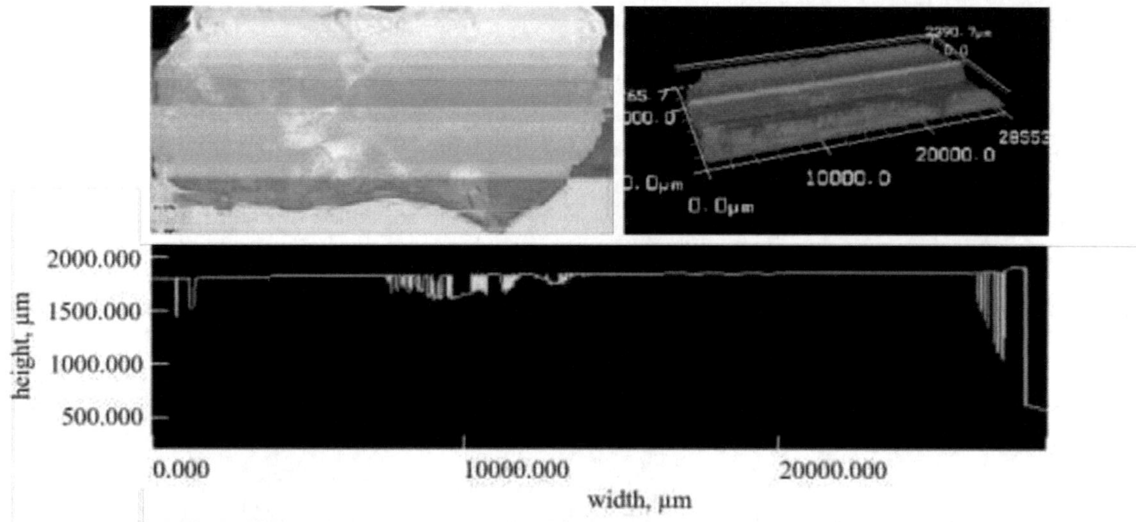

Figure 9. Three-dimensional view and height profile of a slice of porcine lung tissue taken by a laser-scanning microscope of the type Keyence VK-9710, v_f = 1 mm/s, p = 70 MPa

Figure 9 shows the profile of the cut slice. The upper pictures show a three dimensional view of the slice, the bottom picture shows a height profile over the complete width. It can be seen, that the slice is sufficiently plane-parallel with a cragged area on the cut surface. Despite the cragged area, there is enough plane surface for a microscopic view at the tissue. The plane-parallel surface allows further investigations with different optical microscopy techniques with gaining a high depth of field.

To keep the tissue in its vital state, it will be necessary to cultivate the tissue after the preparation. It is of great manner for the cultivation to cut thin slices. The tissue has to be perfused with special fluids to simulate its natural habit and to keep it alive. The goal was to cut slices thinner than 500 µm. The slice shown in Figure 9 has a thickness of about these 500 µm, which fits the requirements already. From this point, we cut increasingly thin slices by maintaining the process parameters and just repositioned the cutting path closer to the vertical low-pressure suction to determine the possible minimum of slice thickness with our process.

By performing further cuts, we managed to lower the slice thickness from 500 µm to below 200 µm. We cut even thinner slices, but not reliably. Within our experiments with porcine lung, failing to cut a slice from the tissue was not a major problem. Considering the poor availability of resected human tissue, the reliability of the cutting process is of paramount importance. Keeping this in mind, we conclude that we excelled our goal of cutting highly plane-parallel soft tissue slices with a thickness of less than 500 µm by managing to reliably cut slices with a thickness of 200 µm.

5 CONCLUSIONS

The goal of this research project was to prove the feasibility of atraumatic cuts of biologic soft tissue by means of high-pressure water jets. We exemplarily used porcine lung tissue for our investigations. Due to its spongy and soft texture, the preparation of lung tissue faces many problems. According to the previous state of science and technology, several compromises had to be made, to be able to cut soft tissue for research purposes. The tissue had to be cut without infusing it with stabilizing media like low-melting agarose. The slices should have a thickness below 500 µm and a high grade of plane-parallelism.

These goals could be fully achieved and even excelled. We built a device for the fixation of soft tissue with little damage using a low-pressure suction. This enabled us to atraumatically prepare porcine lung tissue for scientific investigations. The minimum slice thickness we managed to achieve was about 200 µm. The grade of plane-parallelism was sufficiently high.

As a next step, we want to transfer our results on cuts of other biological tissue like liver or spleen. Furthermore it will be necessary to develop a system that enables to cut within a sterile environment with a sterile medium.

The preparation of high-precision soft tissue transections leads to possibilities of cultivation of native living soft tissue. This creates the basis for observing human or animal bodily mechanisms using optical microscopy techniques in real time. Not relying on examining immortalized tissue or cells outside of its tissue context holds a great potential for major medical advances, such as in autoimmune research.

6 ACKNOWLEDGEMENTS

This study was supported by the German Research Foundation (Deutsche Forschungsgemeinschaft, DFG, Collaborative Research Center SFB-TR84).

7 REFERENCES

[1] Perrin, S. "Preclinical research: Make mouse studies work" *Nature: 507,* 2014; pp. 531-533.

[2] Grizzle, W. E., W. C. Bell, K. C. Sexton. "Issues in collecting, processing and storing human tissues and associated information to support biomedical research." *Cancer Biomark: 9(1-6)*, 2010, pp. 531-549.

[3] Sanderson, M. J. "Exploring lung physiology in health and disease with lung slices." *Pulmonary Pharmacology & Therapeutics: 24(5)*, 2011, pp. 452-465.

[4] Gerckens, M., et al. "Generation of Human 3D Lung Tissue Cultures (3D-LTCs) for Disease Modeling." *Journal of Visualized Experiments,* 2019.

[5] Rosales Gerpe, M. C., et al. "Use of Precision-Cut Lung Slices as an Ex Vivo Tool for Evaluating Viruses and Viral Vectors for Gene and Oncolytic Therapy." *Molecular Therapy - Methods & Clinical Development*, 2018, pp. 245-256.

[6] Biskup, C., et. al. "Heat Generation During Abrasive Water-Jet Osteotomies Measured by Thermocouples." *Strojniški vestnik – Journal of Mechanical Engineering: 52(7-8)*, 2006, pp. 451-457.

[7] Bauer, M., et al. "Geometric adaption of resorbable myocardial stabilizing structures based on the magnesium alloys LA63 and ZEK100 for the support of myocardial grafts on the left ventricle." *Biomedical Engineering: 57,* 2012, pp. 22-25.

[8] Hinte, N., et al. "Stabilizing structures for the aorta replacement – Static and dynamic testing." *Jahrestagung der Deutschen Gesellschaft für biomedizinische Technik (BMT)*, 2010.

2019 WJTA Conference and Expo
November 11-13, 2019 • New Orleans, Louisiana

Paper

SAFETY CONSIDERATIONS FOR LOW PRESSURE CLEANING APPLICATIONS

J. Ashmead
DuPont Co.

D. Wright
StoneAge Inc.

ABSTRACT

Low pressure cleaning is commonly conducted on industrial sites by contractors utilizing "Pressure Washer" pumps and tooling such as guns or wands designed for the pressure washer cleaning market. Site Personal Protective Equipment (PPE) requirements for use while operating this equipment often consist of safety glasses, face shield, gloves, rubber boots and rain suits, as compared to the minimum requirements of metatarsal and shin protection for hand-held waterblast cleaning. Operating pressures up to 27.5 MPa (4000 psi) and flow rates up to 76 liters/minute (20 gallons per minute) are possible within the pressure washer realm. The purpose of this study was to determine the level of jet impact protection provided by typical rubber boots and rain suits, as well as tests on simulated skin to demonstrate the effect of jet impact with various combinations of pressure and flow within the pressure washer cleaning range.

Organized and Sponsored by the WJTA®

1. INTRODUCTION

Contractors and industrial sites frequently use pressure washers for low pressure cleaning applications, such as light foulant removal from valves and flanges. These machines are less expensive to purchase and operate than a typical waterblast pump unit when the jetting power of a larger pump is not necessary. The PPE requirements and gun design requirements for these units are below those of typical waterblast industry requirements that are in place to prevent waterjet injuries to operators. With the higher flow capabilities now available in pressure washer pumps, the combination of a higher flow, even at a lower pressure, can generate substantial jet impact, approaching that of the lower end of what would otherwise be considered waterblast power.

Waterblast handheld shot guns have a required minimum barrel length of 122 cm (48 inches) and an overall length requirement of 168 cm (66 inches) from the nozzle to the shoulder rest. This design physically prevents the operator of the gun from passing the jets across the feet or legs when properly held, preventing accidental jet contact with the body. Guns or wands used within the pressure washer industry are typically much shorter for ease of use.

Minimum PPE requirements for handheld waterblast shotgun work include metatarsal and shin protection for the gun operator, and face shields for protection from debris as well as rain suits for chemical and splash protection. PPE for pressure washing may be as minimal as steel toe rubber boots, rain suits and goggles or face shields.

One potential difference in pressure washer applications is the common use of a fan jet, as compared to a round jet, although round jets may still be used with a pressure washer for difficult, harder deposits. A fan jet quickly dissipates in impact energy with distance compared to a round jet; however, a fan jet at a very close distance can have an impact energy nearly equivalent to a round jet. Rotating "turbo" nozzles are also commonly used, and when rotating properly, they would be expected to have an impact effect similar to a traversing round jet orifice, although with the effect of multiple passes if held in one place on a surface.

2. TEST ARRANGEMENT

Tests were conducted by traversing a jet across samples at a rate of 150 mm/second (6 in/second) at pressures and flow rates of 3.4 to 27.6 MPa and 3.8 to 76 lpm (500 to 4000 psi and 1 to 20 gpm). At the center of the sample, the traversing motion was stopped, and the time required for the stationary, or static, jet to penetrate the sample was recorded. Both round and 15 degree fan jets were compared at several standoff distances from the samples.

The test samples were held in place by an assembly that allowed them to be tensioned, keeping the surface of the samples from being deflected by the jet impact. The simulated skin samples were tanned cowhide, 1 mm thickness, smooth finished on one side, as suggested by the paper titled "Ballistic Skin Simulant"[1] for research in forensic science. Samples of fresh pig skin with a thickness of 4 mm were used for comparison, since this material has also been used in evaluations of jet impact. The rubber boot samples were taken from the upper and ankle section of new steel toe rubber work boots commonly worn in industry. These samples consisted of both single layer and double layer rubber coating on a fabric backing, with combined thicknesses of 1.63 and 2.54 mm (.064 and .100 in). The rain suit samples came from lightweight fire resistant PVC/Polyester rain suit pants with a thickness of 0.38 mm (.015 in). The sample holder and traversing lance with nozzle are shown in Figure 1, and the boots and rain suit types tested are shown in Figure 2.

3. TEST RESULTS

3.1 Round Jets and Skin Simulant

3.1.1 Standoff Distance

Initial tests on the skin simulant were conducted using a 0.8 mm diameter (.031 in) round carbide orifice to determine the effect of standoff distance, with the intention of conducting testing within the range that would be the most likely to result in damage. It was found that within the standoff distance range of 10 to 76 mm (0.38 to 3 in) there was not a significant difference in result for penetration of the simulant, although with increasing standoff distance the width of cut did increase.

3.1.2 Static Jet Penetration

The time required for the static jet to completely penetrate the skin simulant was measured for several pressures, flows from 2.3 to 38 lpm (.6 to 10 gpm) and standoff distances of 10 to 76 mm, but there were no significant differences among these variables other than the slight change in pressure. Using a 0.8 mm (.031 in) round orifice at 3.4 MPa (500 psi) and 2.3 lpm (.6 gpm), the sample was penetrated in 5 seconds, while at 4.8 MPa (700 psi) and 2.6 lpm (.7 gpm), the sample was penetrated in less than 1 second, both tests conducted with a 76 mm (3 in) standoff distance. Essentially, once the threshold pressure of the material was reached, combined with the 1 mm thickness of the material, the penetration time was very short for all flow rates tested.

Tests were conducted on the 4 mm fresh pig skin using the 0.8 mm (.031 in) orifice at the 76 mm (3 in) standoff distance, with the result that at 6.9 MPa (1000 psi), it required 6 seconds to achieve full penetration through the sample by the jet.

3.1.3 Traversing Jet Penetration

The traversing jet tests were conducted with three orifice sizes, 0.8, 2.0 and 3.0 mm (.031, .078 and .118 in) and a higher pressure was used in each individual pass over fresh material surface until the skin simulant was cut through along the traverse, using a standoff distance of 76 mm (3 in). The results of this testing showed a decrease in the pressure required to achieve the cut as the flow rate was increased from 4.5 lpm (1.2 gpm) to 25 lpm (6.5 gpm). The 0.8 mm (.031 in) orifice achieved a cut through along the traverse at 13.8 MPa (2000 psi) with a corresponding flow rate of 4.5 lpm (1.2 gpm) while the 2.0 mm (.078 in) orifice achieved a cut at 11 MPa and 25 lpm (1600 psi and 6.5 gpm). Increasing to the 3.0 mm (.118 in) orifice resulted in a smaller decrease in pressure, requiring 10.3 MPa while flowing 55 lpm (1500 psi and 14.5 gpm). This result is plotted in Figure 3. The higher flow rates at lower pressures produced tearing of the simulant along with

a wider jet impact. Figures 4 and 5 show the resulting cuts produced with the 0.8 mm orifice compared to the 3.0 mm orifice.
Traversing jet tests were also conducted on the pig skin samples with the 0.8 mm (.031 in) orifice at a standoff distance of 76 mm (3 in). A pressure of 20.7 MPa (3000 psi) with a flow rate of 5.3 lpm (1.4 gpm) was reached before penetration began to occur along the jet path.

3.2 Round Jets and PPE Resistance

3.2.1 Rubber Boot Static Jet Penetration

The 0.8 mm (.031 in) orifice was used for a static jet penetration test on a sample from the rubber boot. At a pressure of 6.9 MPa (1000 psi) producing a flow rate of 3 lpm (.8 gpm), the jet penetrated the material in less than 2 seconds at a standoff distance of 38 mm (1.5 in). Essentially, the rubber boot material would not offer protection at close range from a static round jet with pressures at or above 6.9 MPa (1000 psi), well within the range of the smallest pressure washers.

3.2.2 Rubber Boot Traversing Jet Penetration

The traversing jet tests were conducted with four orifice sizes to test for the effect of increasing flow rate on the minimum pressure to produce a penetration completely through the material. All tests were conducted at a standoff distance of 38 mm (1.5 in). The results showed the trend that increasing flow resulted in a lower pressure required to achieve penetration completely through the material. The smallest orifice, 0.8 mm (.031 in), with a flow of 5.7 lpm (1.5 gpm), was able to reach a pressure of 24.1 MPa (3500 psi) before the jet began to penetrate the rubber boot material along the traverse as pinholes. This test was repeated with a sample of skin simulant underneath the boot material to determine if the material was absorbing any of the jet energy; the traversing jet penetrated both the boot and skin simulant, as shown in Figure 6.

When the orifice size was increased to 1.6 mm (.062 in), the boot material was cut completely through on the traverse with 20.7 MPa (3000 psi) at the accompanying flow rate of 21.6 lpm (5.7 gpm); with the 2.0 mm (.078 in) orifice, the necessary pressure to cut reduced to 17.2 MPa (2500 psi) with a flow rate of 31 lpm (8.2 gpm). A 2.5 mm (.100 in) orifice cut through on traverse at a pressure of 13.8 MPa (2000 psi) and a flow rate of 45.4 lpm (12 gpm). These results are shown in the chart in Figure 7. The 0.8 mm and 2.5 mm (.031 and .100 in) orifice tests were repeated at the same pressures on the dual layer boot material, and the increased stiffness of the thicker material resulted in more defined traversing cut results.

3.2.3 Rain Suit Resistance to Round Jets

The static and traversing round jet tests on the plastic coated rain suit material were conducted at a standoff distance of 38 mm (1.5 in). The 0.8 mm (.031 in) orifice required less than 1 second to penetrate at 6.9 MPa (1000 psi) with a flow rate of 3 lpm (.8 gpm) in the static jet test. With this same orifice size, the pressure required an increase to 10.3 MPa (1500 psi) with 3.8 lpm (1 gpm) to produce a continuous cut through this material when traversing. A 1.6 mm (.062 in) round orifice achieved a traversing cut through the rain suit at a pressure of 6.9 MPa (1000 psi) and a flow rate of 12.5 lpm (3.3 gpm). This rain suit material was less resistant to the jet impact than the skin simulant and should not be considered as protection from any jet.

3.3 Fan Jets and Skin Simulant

3.3.1 Static Jet Penetration and Standoff Distance

Fan jets are commonly used with pressure washer systems, and similar tests were conducted with 15 degree fan jets in equivalent orifice sizes to the round jet tests. The effect of increasing standoff distance of the orifice from the material showed a significant influence on the pressure required to penetrate the skin simulant with a stationary or static jet due to the rapid dissipation of jet energy. At the closest standoff distance of 6.4 mm (.25 in), the minimum pressure required for static jet penetration was similar to the round jet results in all except the smallest orifice size (0.9 mm or .036 in), which required a pressure of 10.3 MPa (1500 psi) at 5.3 lpm (1.4) gpm to achieve penetration in 2 seconds or less. The 1.6 and 2.0 mm (.064 and .080 in) fan jets penetrated the skin simulant at 5.5 MPa (800 psi) with flow rates of 11.7 and 18.5 lpm (3.1 and 4.9 gpm) respectively.

The same fan jet orifice sizes were tested at standoff distances of 38 and 76 mm (1.5 and 3.0 in) for complete static jet penetration of the skin simulant in 2 seconds or less. These results are shown in Figure 8 along with the 0.8 mm (.031 in) round jet results. All points plotted would cause significant injury and are well within the range of commonly available pressure washers; this plot does show that a round jet can carry impact energy further than a fan jet.

3.4 Fan Jets and PPE Resistance

3.3.2 Rubber Boot Static Fan Jet Penetration

The results of the static fan jet tests on the rubber boot material showed that with the 0.9 mm (.036 in) equivalent orifice size at a standoff distance of 6.4 mm (.25 in), penetration occurred at 13.8 MPa (2000 psi) with a flow rate of 6 lpm (1.6 gpm). A 2.6 mm (.103 inch) equivalent fan jet orifice penetrated the material at 6.9 MPa (1000 psi) in less than 1 second, with a flow rate of 34 lpm (9 gpm), again showing that this material would not offer protection at close range from static round or fan jet penetration.

3.3.3 Rubber Boot Traversing Fan Jet Penetration

The traversing fan jet tests were conducted at standoff distances of 6.4 to 76 mm (.25 to 3.0 in) with equivalent orifice sizes from 0.9 to 2.6 mm (.036 to .103 in) and the plane formed by the 15 degree fan pattern oriented parallel to the traversing axis to represent a worst case condition. The results at the 6.4 mm (.25 in) standoff distance are shown versus flow rate in Figure 9 and compared to the similar test with round jets; the smallest orifice size produced a cut through on traverse at 24.1 MPa (3500 psi) while the largest orifice cut through at 16.5 MPa (2400 psi). The fan jet results at a 6.4 mm (.25 in) standoff distance are comparable to the round jet results at 38 mm (1.5 inches) standoff distance. The largest fan jet orifice size (2.6 mm or .103 in) was tested with a standoff distance of 76 mm (3.0 in) with traverse at a pressure of 27.6 MPa (4000 psi); this did not produce a cut through the rubber boot material, but once the traverse was stopped, full penetration occurred in less than 1 second.

3.3.4 Rain Suit Resistance to Fan Jets

The static and traversing fan jet tests on the plastic coated rain suit material were conducted at a standoff distance of 6.4 mm (.25 in). The 0.9 mm (.036 in) equivalent fan jet orifice required less than 1 second to penetrate at 6.9 MPa (1000 psi) with a flow rate of 4.2 lpm (1.1 gpm) in the static jet test. With this same orifice size, the pressure required an increase to 11 MPa (1600 psi) with 5.3 lpm (1.4 gpm) to begin to produce a cut through this material when traversing, similar to the round jet tests and again demonstrating that this material does not offer protection from a round or fan jet impact.

4. CONCLUSIONS

The purpose of these tests was to demonstrate the potential risks of jet impact produced in the pressure and flow range of what is commonly considered to be pressure washing and determine the level of jet impact protection provided by common protective wear used in the industry. A significant difference in pressure required to penetrate all samples did exist depending on whether the jet was stationary or traversing the sample. Using larger orifice sizes to increase flow rate resulted in a decrease in the pressure required to produce traversing cuts through the samples. The results also illustrated the greater risk of round jets due to the ability to carry concentrated energy much further than fan jets, although both were capable of penetrating a human skin simulant at close range with pressures as low as 3.4 MPa (500 psi).

The protective equipment tested consisted of rubber boot and plastic coated rain suit samples. The results showed that the rain suit material should not be considered as any level of protection from jet impact, being slightly less resistant than the skin simulant samples. The rubber boot material was able to be penetrated by a stationary jet at 6.9 MPa (1000 psi), and only showed resistance up to 13.8 MPa (2000 psi) with a traversing round jet. This material did show resistance to the traversing fan jet test but only with increased standoff distance; it did not resist close range fan jet impact for either the stationary or traversing jet within the range of what is commonly available in pressure washer systems. At a minimum, operators of these systems should be instructed of the risk of significant injury that can be caused by the jets produced by common pressure washer machines.

REFERENCES

1. Jussila J, Leppaniemi A, Paronen M, Kulomaki E. "Ballistic Skin Simulant". *Forensic Science International.* 2005, Volume 50(1), pages 63-71.

Figure 1. Sample Holder and Traversing Nozzle Lance Test Arrangement

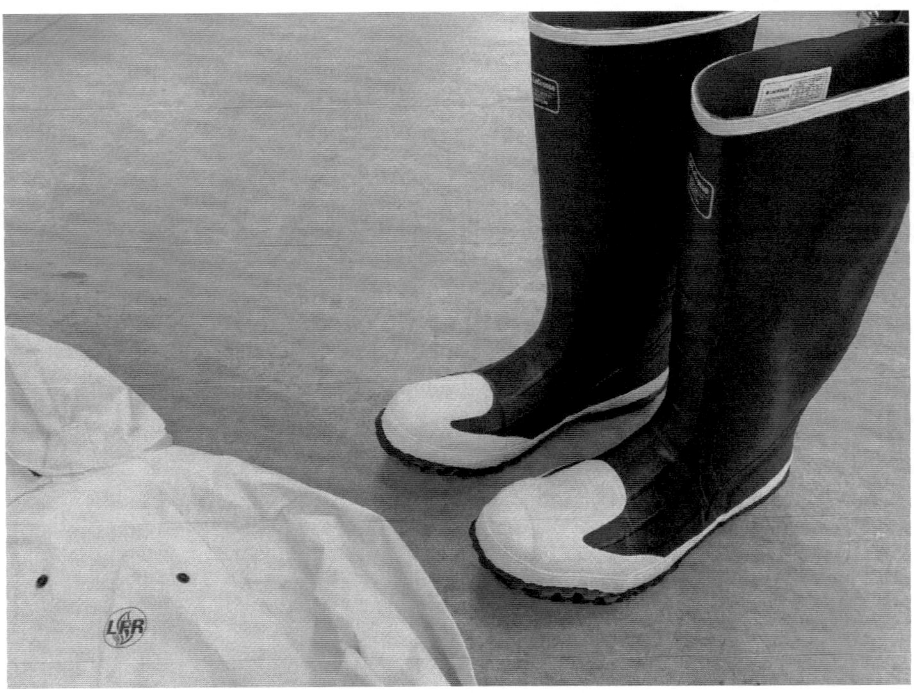

Figure 2. Rubber Boots and Rain Suit of the Types Tested

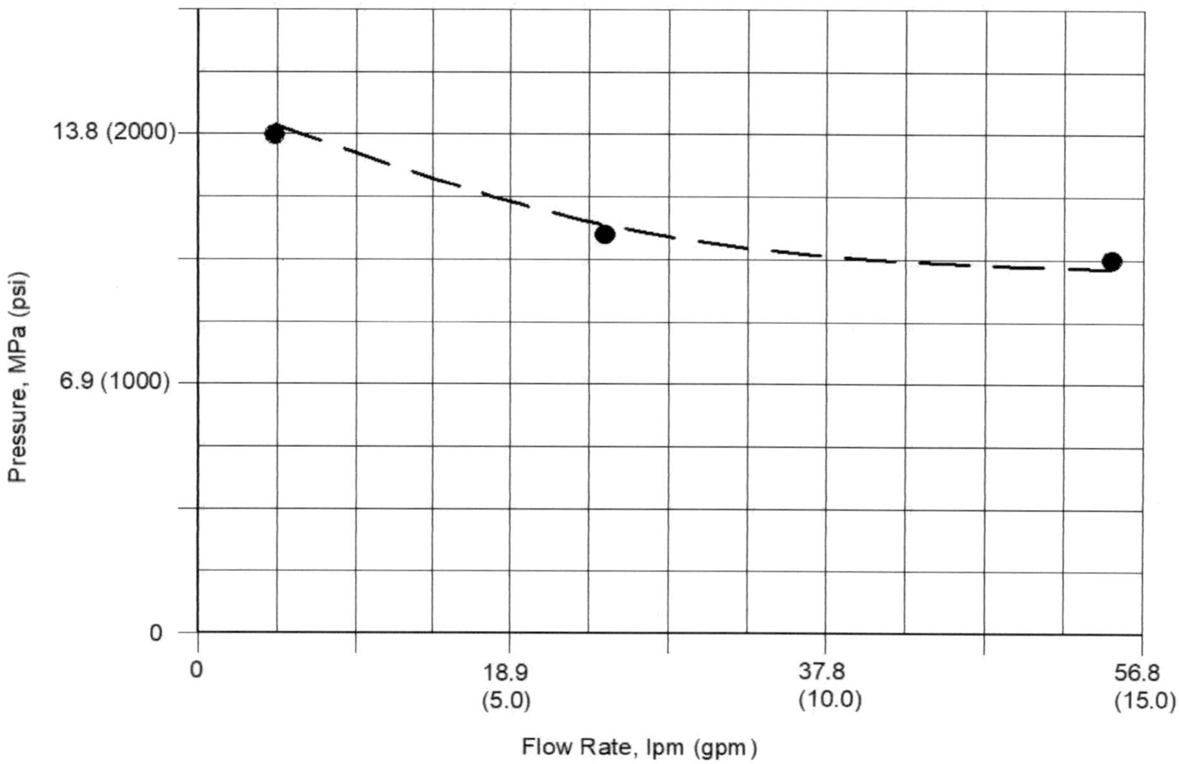

Figure 3. Pressure and Flow Rate to Result in Cut from a Traverse of Skin Simulant

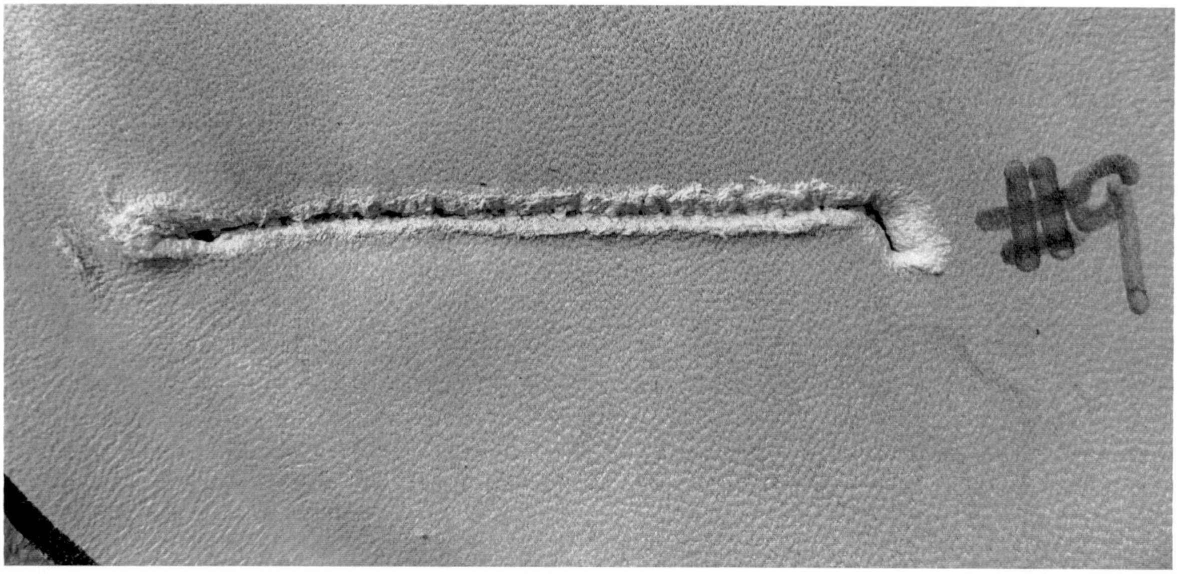

Figure 4. Cut from a Traverse of Skin Simulant by 0.8 mm (.031 in) Orifice at 13.8 MPa and 4.5 lpm (2000 psi and 1.2 gpm)

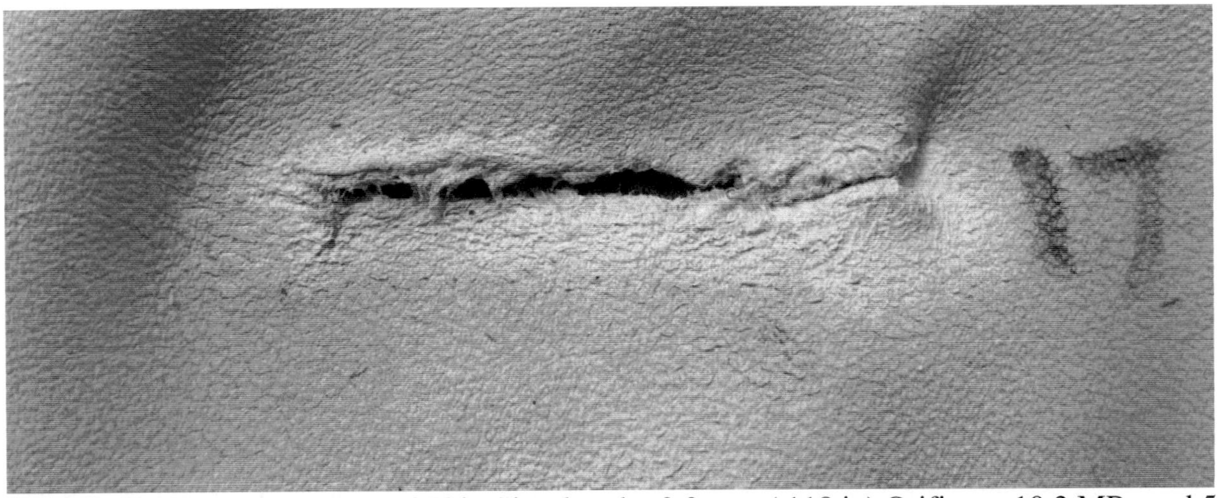

Figure 5. Cut from a Traverse of Skin Simulant by 3.0 mm (.118 in) Orifice at 10.3 MPa and 55 lpm (1500 psi and 14.5 gpm)

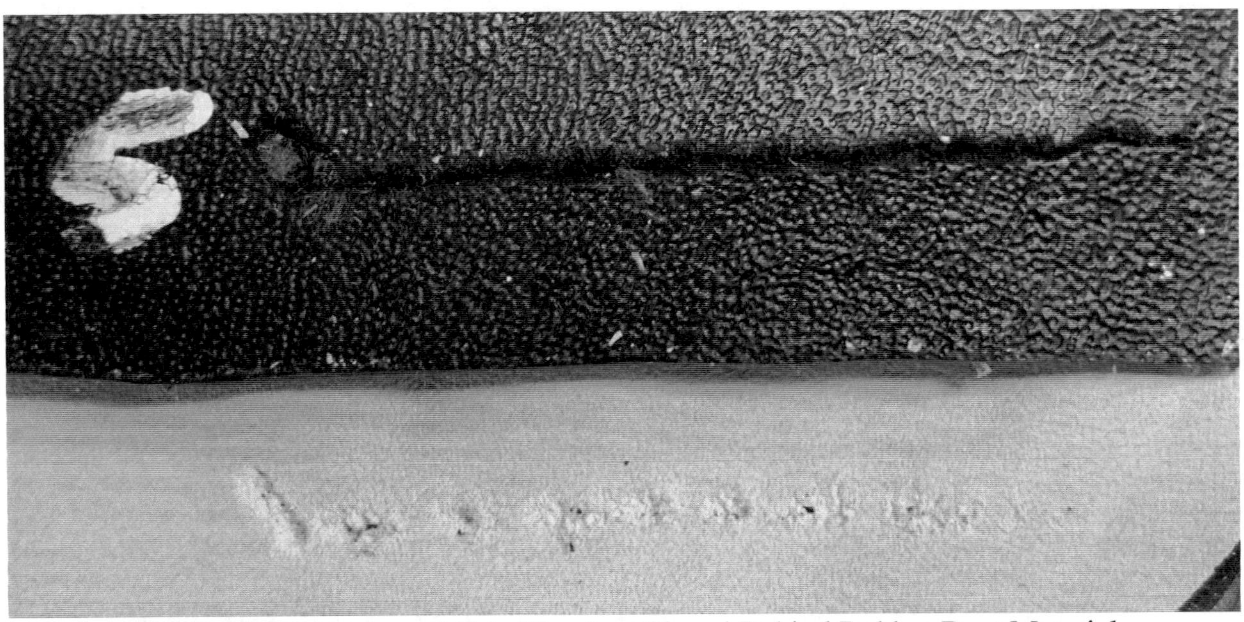

Figure 6. Penetration of Skin Simulant Placed Behind Rubber Boot Material

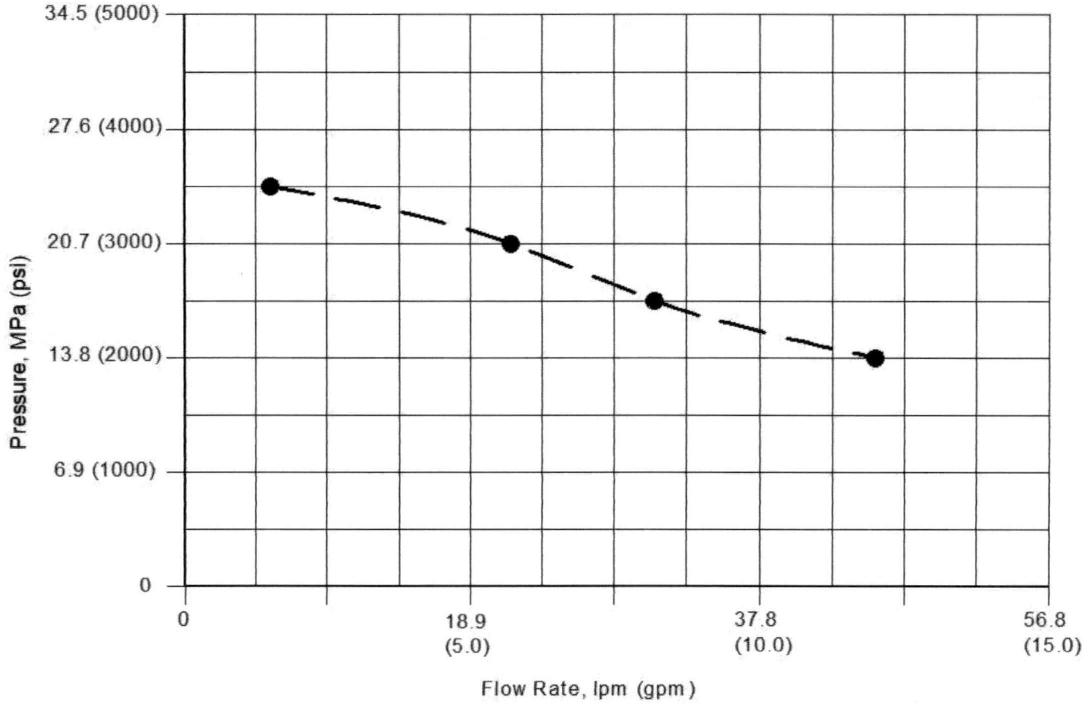

Figure 7. Pressure and Flow Rate to Result in Cut from a Traverse of Rubber Boot Material

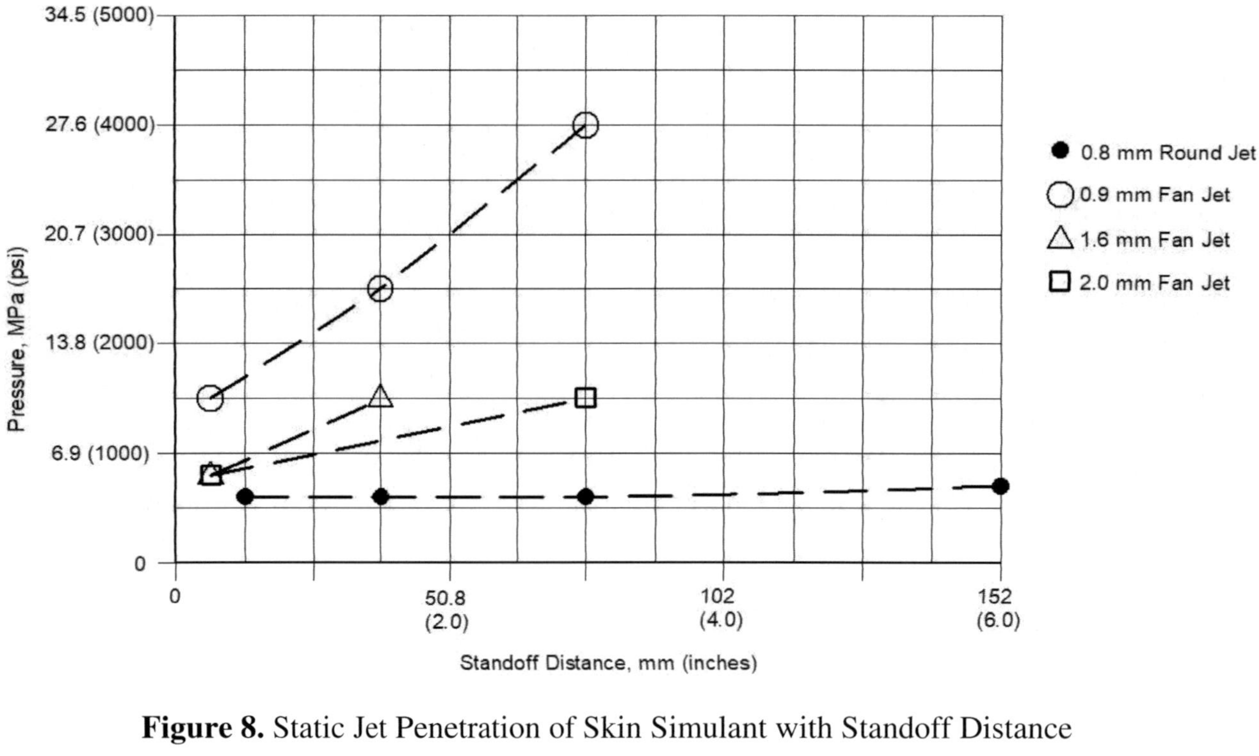

Figure 8. Static Jet Penetration of Skin Simulant with Standoff Distance

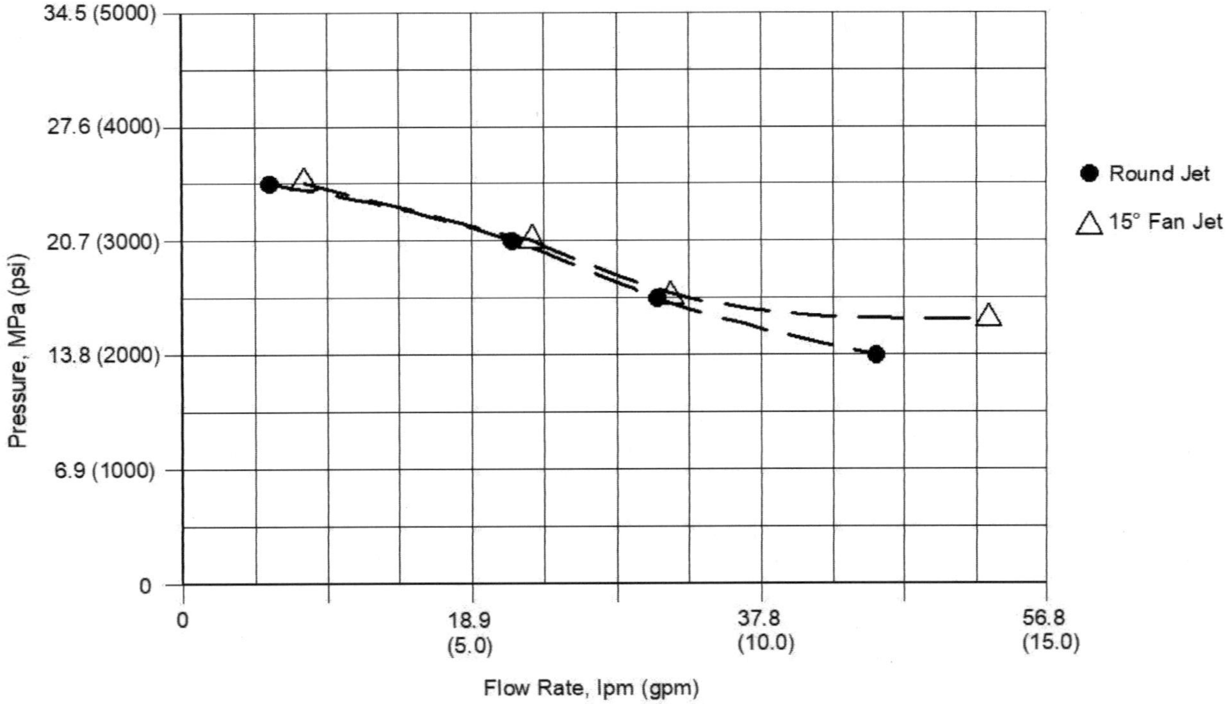

Figure 9. Pressure and Flow Rate to Result in a Cut from a Traverse of Rubber Boot Material

2019 WJTA Conference and Expo

November 11-13, 2019 • New Orleans, Louisiana

Paper

COMPARISON OF SSPC-NACE and ISO WATERJET CLEANING STANDARDS FOR COATINGS REMOVAL

Lydia M. Frenzel
Advisory Council
Vancouver, WA USA

ABSTRACT

The current SSPC-NACE series "Joint Surface Preparation Standard Waterjet Cleaning of Metals" which covers four levels of cleanliness is compared to the ISO 8501-4 "Preparation of steel substrates before application of paints and related products — Visual assessment of surface cleanliness — Part 4: Initial surface conditions, preparation grades and flash rust grades in connection with high-pressure water jetting." ISO 8501-4 revision is underway. The current "Joint SSPC Surface Preparation Standard/ NACE Standard Practice: Wet Abrasive Blast Cleaning" series is compared to the wet abrasive blast section of ISO 8504-2 "Preparation of steel substrates before application of paints and related products — Surface preparation methods — Part 2: Abrasive blast-cleaning" which is in the Final Draft International Standard (FDIS) stage and will be published shortly. Every global contractor should be aware of similarities and differences.

Organized and Sponsored by WJTA

1. INTRODUCTION

Coatings removal and subsequent application of paints and coatings require a depth of knowledge which is beyond industrial cleaning. Lou Vincent estimated that 85% of premature coatings failures occur because of the condition of the substrate upon which the paint is applied. Many waterjet contractors want to "clean" or remove coatings" but to be successful they must learn the language and details which concern the corrosion managers of the ultimate client. If the firms work globally, they have to be familiar with not only the standards of their home country but of global standards which in the coatings industry means ISO standards. Coatings are more than cosmetic; coatings and paints are the first line of defense against corrosion. At first glance, the ANSI, NACE, SSPC and ISO standards seem to be interchangeably, but the devil is in the details. This author has worked on US and ISO standards for years, and still gets surprised in the face-to-face discussions where each country is vying for "their" national body words and emphasis.

2. WATERJET CLEANING

What is the same? SSPC-NACE WJ-1, WJ-2, WJ-3, WJ-4 and ISO 8501-4 describe conditions for four levels of cleanliness, ranging from "removal of loose material" to clean to bare substrate." The abrasive blast cleaned documents came first. Inspectors are trained and certified predominately on the abrasive blast standard language and the abrasive blast photographs. This unconscious comparison with abrasive blast can, and does, lead to disagreements on field sites, and sometimes back in the testing labs.

The ISO 8501-4 was originally written with language from the abrasive cleaned standards and includes all the levels of cleanliness. The SSPC-NACE documents started with completely different language from an earlier "pressure washer" concept and the first edition included four levels of cleanliness in one document; later editions separated SSPC-NACE into four documents.

There is a pretreatment section in standards. In all cases, it is assumed that the surface is free from oil and grease, and that the heavier layers of old rust have been removed, typically by chipping tools. In the case of WJ cleaning, often the pretreatment for removal of oil can be skipped, but then the cleaned surface should be examined for areas of water droplets before paint application.

What is different? The US documents started with the descriptive language, followed with guidance photographs. ISO standards started as visual pictures with little description. Their initial language was simple but based on prior abrasive standards. When it came time to describe the four levels of visual cleanliness, the ISO 8501-4 [REF 1] left off the "highest" level because the author, or project leader, could not see the difference as clear as in the "standard" abrasive photos, and moreover, had left no room in the descriptive language. SSPC-NACE has the Surface Preparation separated into 4 documents. [REF 2-5]

In 2002 as the ISO WJ document was being written, the project leader didn't understand that abrasives clean from the top down, leaving material in pits and crevices; waterjet cleans from the pits and crevices up, leaving material on the top. This effect comes from the hydraulic push between layers. Consequently, the project leader did not accept descriptions or photos of Wa 3.

At the time of publication, the European WJ cleaning systems were predominately 15,000 psi to 20,000 psi (103 MPa to 138 MPa), which is a mix of hydraulic and cavitation actions which led to partial paint layer removal. The USA contractors had quickly moved to the >30,000 PSI (207 MPa) which had cavitation as the major cleaning method to chip away the paint layers to the substrate. The difference in threshold energy between the 20,000 psi (138 MPa) and the 40,000 psi (276 MPa) is critical to removal of coatings layer by layer or stripped down to the metal substrate. The Project Leader for ISO 8501-4 didn't think, in his experience, that waterjetting equipment could achieve the highest level of cleanliness, that is surface preparation which promotes top performance. The coatings manufacturers, notably International Paint and Hempel who did the initial research, ascertained that the performance was of highest quality. International Paint, in 1994, advocated for WJ cleaning, particularly for marine applications where invisible salt was causing corrosion problems so that ships broke up at sea. The loss of ships led to coatings in ballast tanks being designated a primary requirement for Safety Of Lives At Sea (SOLAS) regulations.

STG Guide No. 2222 is the first published reference on high-pressure WJ use in coatings removal. The photos clearly show an underlying black iron oxide layer on the metal which is not removed. [REF 6] This set of reference photo verify that water permeates or diffuses over time through the paint and forms an inert black oxide layer at the substrate. STG Guide No. 2222 was a hindrance to accepting WJ as a cleaning technique, because the conventional wisdom was to remove coating to the bare "gray" steel coloration, not leave old rust layer on the substrate. Never mind that this oxide layer formed under the coating over time and was not the root cause of any coating failure. IN STG No. 2222, only a field specimen which had not been installed on a ship, but had been in the yard, exhibited gray metal when the coating was removed. STG Guide No. 2222 reinforced the Project Leader's idea that WJ couldn't clean to bare substrate.

Thus ISO 8501-4 was published (2006) with no representative photograph of Wa 3, and a definition table which included Wa 1, Wa 2, and Wa 2 ½ and only the following wording for Wa 3:

"NOTE: This part of ISO 8501 does not imply that cleanliness is limited to Wa 2½, but achieving a greater degree of cleanliness could involve a disproportionate increase in time." [REF 1 clause 5 Preparation Grades]

The origin of comparison of "abrasive cleaned" substrates with "WJ cleaned substrates" is important because inspectors make decisions on VISUAL observation. The NACE and SSPC standards started as words to describe the end product based on pressurized water cleaning. The photographs were guides. The WJ standards started with very different descriptions and have gradually moved towards the more established language for the abrasive cleaning. This move towards "abrasive" language makes the unexperienced inspectors think that the WJ clean visual will be the same as the abrasive clean visual. ISO and SSPC-NACE have different titles for the visual levels of cleanliness to emphasize that WJ methods are different from abrasive methodology.

Table 1 Comparison of the Titles and Numbering of SSPC-NACE to ISO standards

Cleaning (abrasive blast)	SSPC/NACE #	Cleaning (water alone)	SSPC/NACE	ISO 8501-4 (2002)
White metal	SP 5/ SP 1	Bare substrate	WJ- 1	
Near White metal	SP 10/SP 2	Very thorough	WJ-2	Wa 2 1/2
Commercial Industrial	SP 6/SP 3 SP 14/SP 8	Thorough	WJ-3	Wa 2
Brush Off	SP 7/SP 4	Light	WJ-4	Wa 1

The language in the ISO had to be revised to include Wa 3. When talking to a group of people who might be using either ISO of SSPC-NACE documents, use the title as the numbers can be confusing.

Current Language in ISO 8501-4 (2006)

Table 2 — Descriptions of the surface appearance after cleaning

Wa 1	Light high-pressure water jetting When viewed without magnification, the surface shall be free from visible oil and grease, loose or defective paint, loose rust and other foreign matter. Any residual contamination shall be randomly dispersed and firmly adherent.
Wa 2	Thorough high-pressure water jetting When viewed without magnification, the surface shall be free from visible oil, grease and dirt and most of the rust, previous paint coatings and other foreign matter. Any residual contamination shall be randomly dispersed and can consist of firmly adherent coatings, firmly adherent foreign matter and stains of previously existent rust.
Wa 2½	Very thorough high-pressure water jetting When viewed without magnification, the surface shall be free from all visible rust, oil, grease, dirt, previous paint coatings and, except for slight traces, all other foreign matter. Discoloration of the surface can be present where the original coating was not intact. The grey or brown/black discoloration observed on pitted and corroded steel cannot be removed by further water jetting.

Table 3 is the **proposed** language which is being balloted currently and will undergo editorial revision. The revised language has to be adopted by a consensus process of about 30 countries.

Table 3 — PROPOSED Descriptions of the surface appearance after cleaning

Wa 1	**Light water jetting** When viewed without magnification, the surface shall be free from visible oil, grease and dirt, loose or defective paint coatings, loose rust and other foreign matter. Any residual contamination shall be dispersed and can consist of firmly adherent coatings, firmly adherent foreign matter and firmly adherent rust.
Wa 2	**Thorough water jetting** When viewed without magnification, the surface shall be free from visible oil, grease and dirt, and most of the rust, previous paint coatings and other foreign matter. Any residual contamination shall be dispersed and can consist of firmly adherent coatings, firmly adherent other foreign matter and firmly adherent traces of rust.
Wa 2 ½	**Very thorough water jetting** When viewed without magnification, the surface shall be free from all visible rust, oil, grease and dirt. Slight traces of firmly adherent thin previous rust; slight traces of firmly adherent thin paint coatings and slight traces of other foreign matter can remain. Any residual contamination shall be dispersed. Discoloration of the surface can be present where the original coating was not intact.
Wa 3	**Water jetting to bare substrate** When viewed without magnification, the surface shall be free from all visible previous rust, oil, grease, dirt, previous paint coatings, and from all other foreign matter. The steel surface might, or might not, appear uniform. Discoloration of the surface can be present where the original coating was not intact.

NOTE 1 Acceptable variations in appearance that do not affect the degree of surface cleanliness defined in the above table include variations caused by composition of the metallic substrate, original surface condition, thickness of the metal, weld metal, mill or fabrication marks, heat treating, heat-affected zones, and differences resulting from the initial abrasive blast cleaning abrasives or the abrasive blast pattern if previously blast cleaned, or waterjet cleaning pattern.

NOTE 2 The grey or brown/black discoloration observed on pitted and corroded steel cannot be removed by further water jetting.

Mill scale is treated as foreign matter. Specks of mill scale under existing coatings will be revealed. Older structures might have intact mill scale under the coatings. Foreign matter and mill scale might be present on Wa 1, Wa 2, Wa 2 ½, but is removed for Wa 3.

What new photos are coming up in the ISO 8501-4 document? What change is occurring now? The ISO committees are adopting a new set of photos and changing the language so that the highest level of visual cleanliness can be included. Photos for level Wa 3 are included, and the new photos for Grade D and Inorganic Zinc are being adopted. As a project leader (PL), this author originally was working with the Working Group (WG) convener at the Swedish Institute of Standards who had not seen or used WJ, but who had experience with abrasive cleaning. This is not an unusual occurrence, even though WJ had been around for over 30 years. It is challenging to educate someone over the internet.

The photographs had to be considered and selected. The original photographs of painted surfaces were part of a Hempel publication. [REF 7] Some of the series just needed to have the Wa 3 (SSPC-NACE WJ-1) photo added. The panels are depicted in a 1:1 scale.

Figure 1 Deteriorated multi-layer coating DCA to Very Thorough Cleaning ISO Wa 2 ½ SSPC-NACE WJ 2

Figure 2 Deteriorated multi-layer coating DCA to Bare Substrate Cleaning ISO Wa 3 SSPC-NACE WJ 1

Wa 2 ½ reveals specks of black mill scale (center of photograph) on the substrate which remained from the prior construction; the substrate is not visible initially, or at preparation grade Wa 1. DCA Wa 3 illustrates the removal of the specks of mill scale. The anchor pattern or profile under the mill scale specks is different from the anchor pattern or profile of the surrounding substrate; the difference can be observed.

Deteriorated Coating B is a heavy layer which came from the Hempel publication. [REF 7]

Figure 3 Deteriorated coating B to Very Thorough Cleaning ISO Wa 2 1/2 SSPC-NACE WJ 2

Figure 4 Deteriorated coating B to Bare Substrate Cleaning ISO Wa 3 SSPC-NACE WJ 1

Figure 3 still has remnants of old corrosion rust and paint on the substrate WA 2 ½ ; the remnants are removed in Figure 4 Wa 3.

The entire series for Condition Grade C steel was replaced. The original Grade C photos had come from Hempel, but the Hempel photo for Wa 2 ½ was a different plate from the Wa 1, Wa 2, and Wa 3 guidance photos. The Swedish convener selected a different photo set which is not depicted in this paper.

It is difficult to photograph, without magnification, the difference between a surface where almost all of the old original rust is removed, and the substrate profile or texture and color is different from the surrounding areas. The absence/presence can be seen in field conditions, particularly if the observer can scratch the substrate. The observer just can't help himself to think that the profile or texture should move to uniformity because that is what his field experience has taught him to expect from abrasive blast cleaning.

The entire series for Inorganic zinc was replaced.

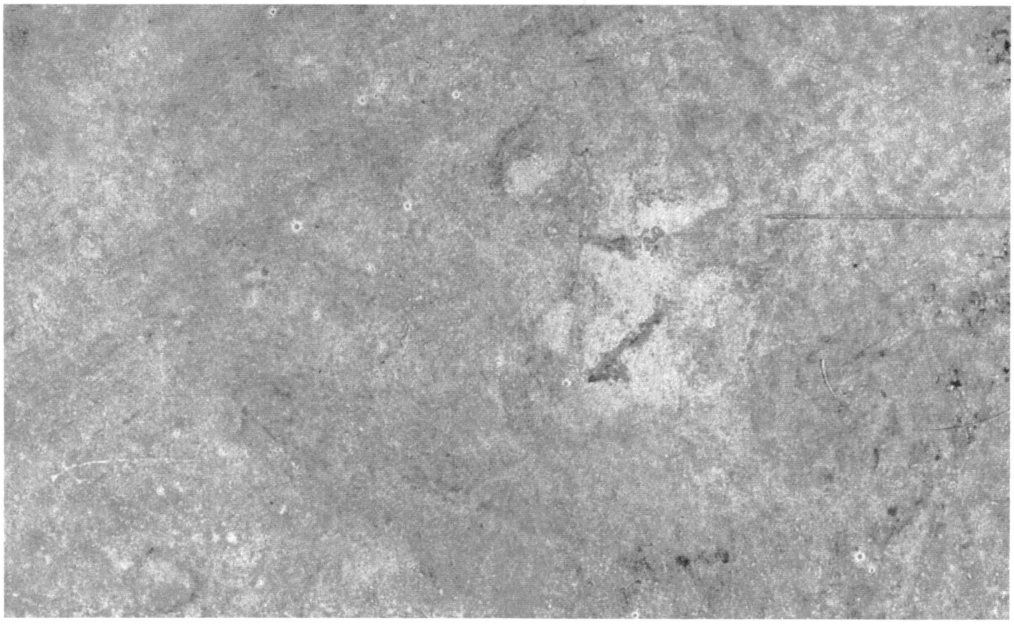

Figure 5 Original Inorganic Zinc

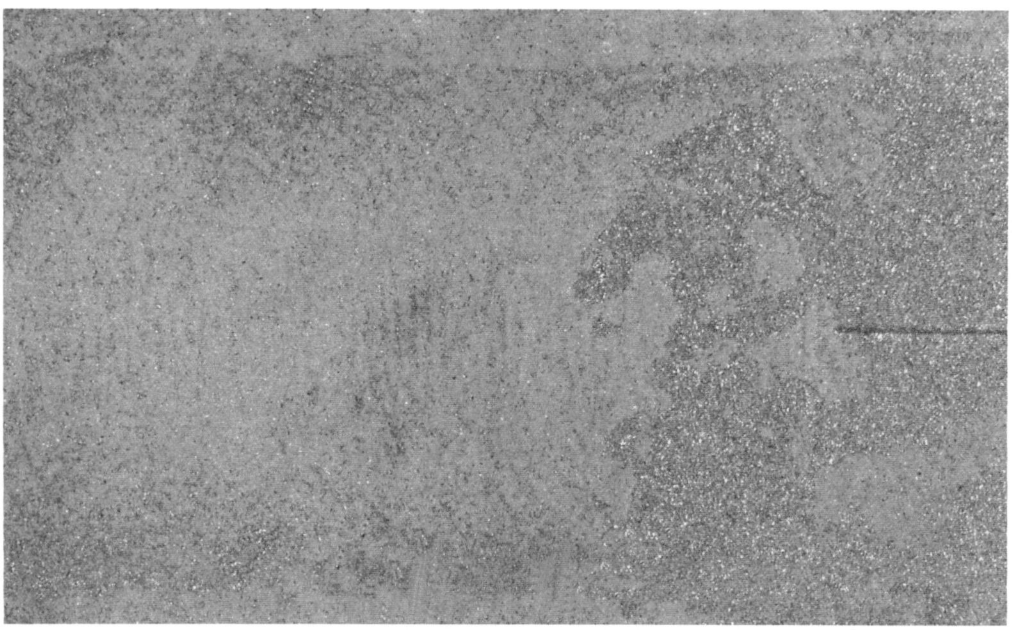

Figure 6 Light Inorganic Zinc cleaned to ISO Wa 1 or SSPC-NACE WJ 4

Figure 7 Thorough Inorganic Zinc cleaned to ISO Wa 2 or SSPC-NACE WJ 3

Here the coating is much thinner than in Figure 6. The lighter-grey blue is the coating; a darker substrate is being revealed.

Figure 8 Very Thorough Inorganic Zinc cleaned to ISO Wa 2 1/2 or SSPC-NACE WJ 2

Here the light blue coating is in small specks over the substrate with the darker substrate apparent.

Figure 9 Bare Substrate Inorganic Zinc cleaned to ISO Wa 3 or SSPC-NACE WJ 1

The zinc coating is off, but the variation in the texture leads the observer to think that something is still on the surface.

It is very difficult to photograph gray on gray. The author has concluded the zinc becomes part of the substrate in the same manner as a phosphate coating becomes part of the substrate. Inorganic zinc is not a polymer coating. With UHP WJ there is no removal benefit from the sideways hydraulic thrust because there isn't a coherent layer. The WJ droplet has to hit every part of the substrate. There are "experts" who say that WJ can't remove zinc completely. However, not removing all of the zinc is true also for abrasive blasting.

With the presence of water, the appearance of flash rust can emphasize the spots where the zinc is not removed because the bare substrate turns slightly brown. Some experts apply a copper ion mist on the substrate which will turn color on steel, but not on zinc. Then they have to remove the copper coat. The experts don't seem to apply the same stringent control on abrasive blast cleaning. It is an unconscious effort to discredit WJ cleaning methodology as compared to abrasive cleaning methodology. For years, the author has looked for any failures where WJ was used to remove zinc with another coating being applied and found none.

The SSPC-NACE WJ standards have moved towards the same sections and language as the traditional abrasive blasting standards. This provides familiarity to the contractor, but also leads to misinterpretation. For example, the pre-cleaning to remove oil and grease is a separate step for dry abrasive blasting. When using UHP WJ, this step often is optional, as UHP WJ removes the oil and grease. HP WJ could or could not remove the oil and grease. Contractors must look and inspect the surface for water droplets as compared to a water sheet.

3. WET ABRASIVE BLAST CLEANING

The wet abrasive blast cleaning (WAB) has its own set of SSPC-NACE joint standards. [REF 8-12] The text sections and wordings are very similar to dry abrasive standards, but they cover both "flash rust" and "rust-back."

The Wet Abrasive Blast Cleaning methodology is part of ISO 8504-2. [REF 13]

Clause 5.2.2.1 **Principle** describes the method for Moisture-injection abrasive blast-cleaning (compressed-air moisture-injection abrasive blast- cleaning).

"This method is similar to compressed-air abrasive blast-cleaning (5.1.2) but with the addition, upstream of the nozzle, of a very small amount of liquid (usually clean, fresh water) to the air/abrasive stream, resulting in a blast- cleaning procedure which is dust-free in the suspended-particle size range of less than 50 μm. The consumption of water can be controlled and is usually 15 l/h to 25 l/h.

Clause 5.2.3.1 **Principle** describes the method for Compressed-air wet abrasive blast-cleaning.

"This method is similar to compressed-air abrasive blast-cleaning (5.1.2) but with the addition of liquid (generally clean, fresh water) downstream of the nozzle to produce a stream of air, water and abrasive."

Clause 5.2.4.1 **Principle** describes the method for Slurry blast-cleaning.

"A dispersion of fine abrasive in water or another liquid is directed, with pumps or compressed air, on to the surface to be cleaned."

There are no separate ISO photos nor definitions, just referral to the ISO 8501-1 photos which were made around 1967 with sand. [REF 14] This can lead to confusion in the field because the appearance of wet abrasive blast cleaned substrates is distinctly darker.

4. CONCLUSION

The wording and guidance photographs for the ISO and SSPC-NACE standard series constantly evolve and are merging together. The primary difference is that SSPC-NACE uses percentages to define what is removed from the substrate, while ISO uses language to generally describe traces of material.

ISO experts recognize that the inclusion of the highest level of cleanliness is achievable for WJ cleaning.

The newer photos give a clearer picture for achievable results, since guidance photographs are critical in the implementation of ISO standards.

Both SSPC-NACE and ISO implementation of WJ text are hindered by adopting language parallel to dry abrasive blast cleaning documents. The similarity of language leads the inspector or assessor to expect similar visual results, when the visual appearance is quite different.

The approach to obtaining consensus language for ISO involves the cooperation of around 30 countries, each with a distinct national interest. The approach to obtaining consensus language for SSPC-NACE standards ultimately involves around 100 individuals, each with a corporate interest.

5. ACKNOWLEDGEMENTS

Thank you to the sponsors of the Advisory Council, and to VLN Technology, Carolina Supply and Equipment, HoldTight, International Surface Preparation, LLC, TermaRust Technologies, and SSPC. NACE Int. has been very supportive during the past three years for ISO activities.

6. REFERENCES

1 ISO 8501-4 (2006) Preparation of steel substrates before application of paints and related products — Visual assessment of surface cleanliness — Part 4: Initial surface conditions, preparation grades and flash rust grades in connection with high-pressure water jetting

2 Joint Surface Preparation Standard Waterjet Cleaning of Metals—Clean to Bare Substrate (WJ-1)

3 Joint Surface Preparation Standard Waterjet Cleaning of Metals—Very Thorough Cleaning (WJ-2)

4 Joint Surface Preparation Standard Waterjet Cleaning of Metals—Thorough Cleaning (WJ-3

5 Joint Surface Preparation Standard Waterjet Cleaning of Metals—Light Cleaning (WJ-4)

6 Schiffbautechnische Gesellschaft e.V., STG Guide No. 2222, Definition of Preparation Grades for high-pressure water-jetting – without addition of solid abrasives, – of corroded and coated steel surfaces, – at different initial conditions ,1st Edition (1992), Hamburg. Can be purchased directly

7 Photo Reference Water Jetting, HMP STD WJ Photo 01 97, Hempel's Photo Reference for Steel Surfaces Cleaned by Water Jetting, issued by Hempel, Copenhagen, 150 Lundtoftevej, 2800 Lyngby, Denmark, January 1997 [this is out of print]

8 Joint SSPC Surface Preparation Standard/ NACE Standard Practice "SSPC-SP 5 (WAB)/NACE WAB-1 White Metal Wet Abrasive Blast Cleaning

9 Joint SSPC Surface Preparation Standard/ NACE Standard Practice "SSPC-SP 10 (WAB)/NACE WAB-2 Near-White Metal Wet Abrasive Blast Cleaning

10 Joint SSPC Surface Preparation Standard/ NACE Standard Practice "SSPC-SP 6 (WAB)/NACE WAB-3 Commercial Wet Abrasive Blast Cleaning

11 Joint SSPC Surface Preparation Standard/ NACE Standard Practice "SSPC-SP 7 (WAB)/NACE WAB-4 Brush-Off Wet Abrasive Blast Cleaning

12 Joint SSPC Surface Preparation Standard/ NACE Standard Practice "SSPC-SP 14 (WAB)/NACE WAB-8 Industrial Wet Abrasive Blast Cleaning

13 ISO DIS 8504-2 Preparation of Steel substrates before application of paints and related products – Surface preparation methods – Part 2: Abrasive blast- cleaning

14 ISO 8501-1 Preparation of Steel substrates before application of paints and related products – Visual Assessment of surface cleanliness – Part 1: Rust grades and preparation grades of uncoated steel substrates and of steel substrates after overall removal of previous coatings

2019 WJTA Conference and Expo November 11-13, 2019 • New Orleans, Louisiana

Paper

JET IMPACT AT CONSTANT POWER WITH THE VARIATION OF PRESSURE AND FLOW AT LARGE STANDOFF DISTANCES

D. Wright
StoneAge, Inc.
Durango, Colorado, U.S.A.

ABSTRACT

In a waterblast system, power is directly proportional to pressure and flow rate; if the pressure is doubled the flow rate is halved for the same power. There is a general knowledge in the industry that the combination of a lower pressure and higher flow can be more effective than the equivalent higher pressure and lower flow combination when applied at the large jet standoff distances typical of vessel cleaning. The purpose of this research was to determine and express in practical terms the effect of the combination of pressure and flow on creating the greatest impact with large standoff distances.

Organized and Sponsored by the WJTA®

1. INTRODUCTION

Standoff distances of 2 meters and greater can be experienced in large vessel cleaning such as reactors, boilers, and storage tanks, where access is limited to a few openings and waterblast systems must be quickly and easily put in place through manways or nozzle openings in the vessel without confined space entry whenever possible. There is a general assumption based on practical experience that effective cleaning at these large jet standoff distances is best accomplished with the combination of lower pressure and higher flow than the equivalent combination of higher pressure and lower flow for the same pump power.

Jet power is directly proportional to pressure and flow; this means that a 450 kW (600 hp) pump could be used to provide 340 lpm (90 gpm) at 69 MPa (10,000 psi) or 170 lpm (45 gpm) at 138 MPa (20,000 psi) as an example. The higher pressure produces a higher jet velocity which is often necessary for cutting into a material to be removed. The higher pressure also requires a smaller orifice size to pass the lower flow at the higher pressure, while a lower pressure requires a larger orifice size to pass the higher flow for the same power. The rate at which a jet deteriorates with distance traveled through air or water is inversely proportional to the orifice size; a jet produced by a smaller orifice will have deteriorated more than a jet produced by a larger orifice at the same distance traveled through the air. This rate of deterioration is assumed to be based only on orifice size, independent of pressure. However, the jets produced by higher pressure are traveling at a higher velocity, meaning a greater relative velocity to the surrounding static air and potentially resulting in faster disruption of the jet, particularly at these larger standoff distances. The jet produced by at 138 MPa is traveling 1.4 times faster than a jet produced by 69 MPa, and a jet produced by 276 MPa (40,000 psi) is traveling twice as fast.

The jet impact force is directly proportional to the mass flow rate of the water, and to the square root of the pressure. Therefore, the lower pressure and higher flow combination will always have more impact force for the same power than a higher pressure at a lower flow combination, although impact force alone does not directly represent material removal capability as velocity is necessary as well. The required jet impact force and jet velocity for successful cleaning varies widely based on the material to be removed and how well bonded it may be to the vessel or structure. In hard materials, such as refractory that is anchored to the vessel wall, higher jet velocities are needed to successfully remove the material, and the standoff distance must be kept to a minimum to maintain the effective velocity. Softer or not well bonded materials may be removed at much greater distances as a different combination of jet impact force and jet velocity are ideal.

The goal of this testing was to determine if a quantifiable difference exists in the rate of jet deterioration at large standoff distances when comparing jets produced by equal power at pressures of 69, 138 and 276 MPa (10,000, 20,000 and 40,000 psi) that would support the assumption that effective cleaning at these large distances can be more effectively accomplished with a lower pressure and higher flow system.

2. TEST ARRANGEMENT

The two types of data collected in this testing consisted of measuring jet impact force on a steel plate with a load cell behind, and depths of cut in a medium-strength, fine grained sandstone block produced by a jet traversed across the face in a single pass to produce a relative measurement of effective remaining jet velocity, as a jet will erode sandstone as the mechanism of cutting. Measurements were taken at standoff distances of 1, 2 and 3 meters from the nozzle exit. Carbide nozzle assemblies were used for the 69 and 138 MPa tests, and large sapphire nozzle assemblies were used for the 276 MPa tests. All nozzles were placed on the end of a .9 m straight feeder pipe to provide good upstream conditions to the nozzle. The tests were performed at powers of 60, 130 and 242 kW (80, 175 and 325 hp) through a single nozzle orifice at each pressure.

3. TEST RESULTS AND ANALYSIS

3.1 Impact Force

The results of the jet impact force tests are shown in Figures 1-3 for the three powers tested at each pressure versus the actual standoff distance. The 69 MPa curves show the least degradation with standoff distance, while the 138 and 276 MPa curves show progressively more loss in impact. The displacement of the power curves on the vertical axis is likely a function of the individual jet quality, depending on the nozzle orifice used in the test, as it would be expected that the curves would otherwise be equally spaced. Figure 4 shows the average impact at each pressure as a percentage of the maximum force measurement for these tests. The greatest impact forces occur at the lower pressure, as should happen for the same power since the impact force is a function of the mass flow rate times the velocity, and velocity is proportional to the square root of the pressure. The 69 MPa result shows a loss of only 8 percent from 1 to 3 meters, while the 138 MPa result show a 20 percent decrease, and the 276 MPa result has a 35 percent loss, showing a pattern of deteriorating impact force with increasing pressure over these standoff distances.

3.2 Depth of Cut

The comparative depth of cut tests in the sandstone block were conducted to reflect the jet velocity, although the mass flow rate increase at each pressure with increasing power also affected the cut depth. In these tests, the maximum depth of cut was achieved at 138 MPa at the 1 meter standoff distance, although the performance of the 138 MPa jet was surpassed at the 2 meter standoff distance and beyond by the 69 MPa pressure at each power tested. The 276 MPa test at the maximum power had a very small erosion of the surface at 2 meters. Figure 5 shows the average depth of cut for the powers tested at each pressure as a percentage of the maximum occurring with 138 MPa at the 1 meter standoff distance, which on average outperformed the 69 MPa tests by 14 percent. However, at the 3 meter standoff distance, the 69 MPa tests exceeded the 138 MPa depths by 10 percent, showing more remaining jet energy.

Figure 6 compares the 69 MPa to the 138 MPa results at the lower test power of 60 kW; the higher pressure being 8% better at 1 meter standoff and the lower pressure better by 9 percent at 2 meters. At 3 meters, the 69 MPa still showed 4 percent of the relative maximum, while the 138 MPa had no effect. Figure 7 shows the relative performance between these pressures at the higher power of

242 kW, where the 138 MPa again outperformed the 69 MPa by 9 percent at the 1 meter standoff, while at 3 meters the 69 MPa result was 21 percent greater than the 138 MPa result. The increasing performance of the lower pressure over the higher pressure at standoff distances of 2 and 3 meters with increasing power shown in comparing these two graphs would indicate that the trend would continue to further benefit the lower pressure with increasing powers at larger standoff distances.

3.3 Combined Effect of Impact Force and Velocity

When waterblast cleaning in vessels and tanks with material deposits such as petroleum coke that may be hard but fractured and not well bonded to the steel structure, it is likely that a combined effect of the impact force of the water combined with sufficient velocity is responsible for removing these materials at a much greater standoff than when the same material is well bonded to a refractory lined surface. This would also likely apply to much softer deposits that are easily removed. To illustrate the effect this would have, the impact force and the depth of cut were multiplied, and the results shown in Figure 8 as a relative percentage. This would emphasize the lower pressure of 69 MPa by 35 to 45 percent over the 138 MPa pressure.

4. CONCLUSIONS

The hypothesis exists that for the same power, the most effective cleaning at large standoff distances in tanks and vessels occurs when utilizing a lower pressure and higher flow combination. This testing illustrated that the impact force will always be greater with the lower pressure, and with increasing standoff distance the velocity of the jet at lower pressure surpasses that of the jet at higher pressure by not losing energy as quickly. The impact force of a 69 MPa (10,000 psi) jet will be 30 to 50 percent greater than the impact force of a 138 MPa (20,000 psi) jet in the standoff distance range of 1 to 3 meters, and the percentage of loss over this distance is greater with the higher pressure. The velocity as based on depth of cut in the sandstone sample showed that the 138 MPa jet would outperform the 69 MPa jet at distances of 1 meter and less but fall below the performance of the lower pressure jet beyond this distance by 10 to 20 percent. If a combined effect of velocity and impact force is compared, the 69 MPa jet could potentially show 35 to 45 percent benefit over the 138 MPa jet at these large standoff distances, where softer or not well bonded brittle materials are being removed. The more rapid loss in relative energy of the higher pressure jet also illustrates why it is important to maintain close standoff distances when using the higher pressures required to remove harder materials such as refractory linings.

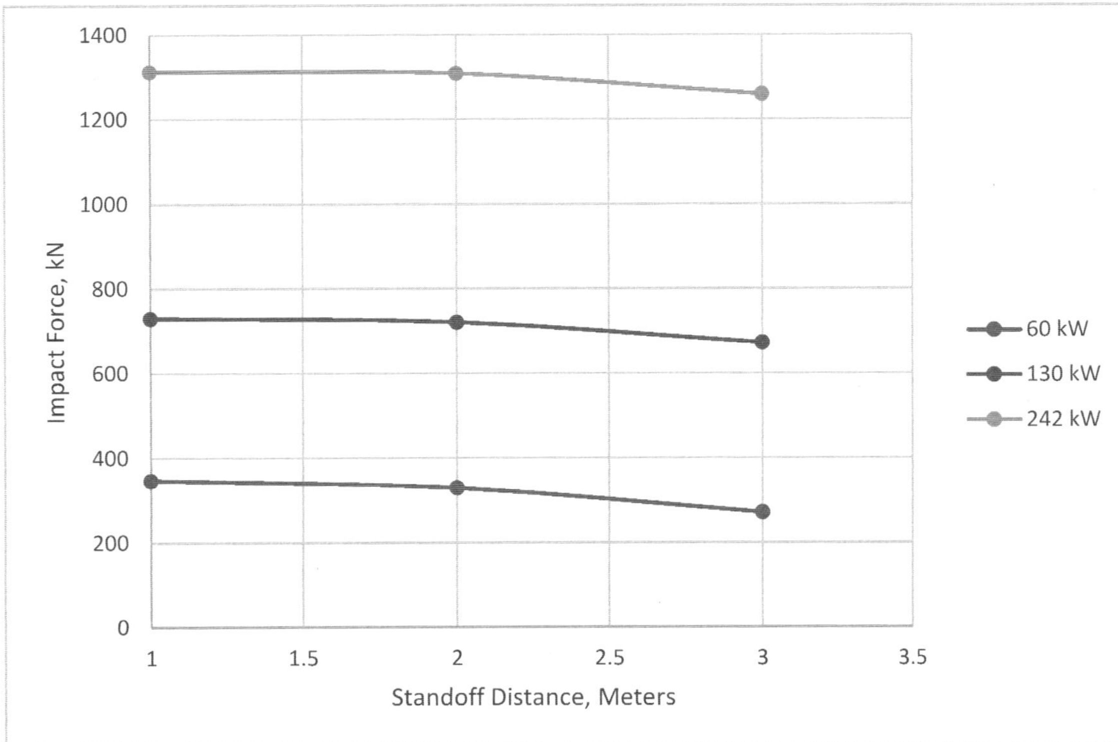

Impact Force by 69 MPa (10,000 psi) at Three Different Powers
Figure 1.

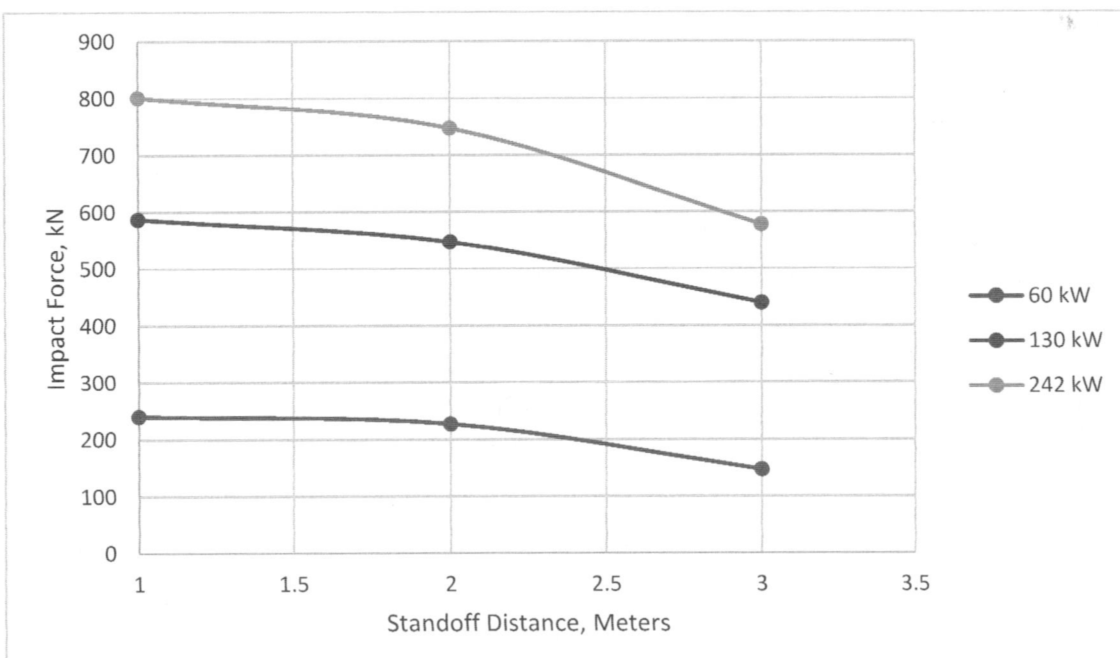

Impact Force by 138 MPa (20,000 psi) at Three Different Powers
Figure 2.

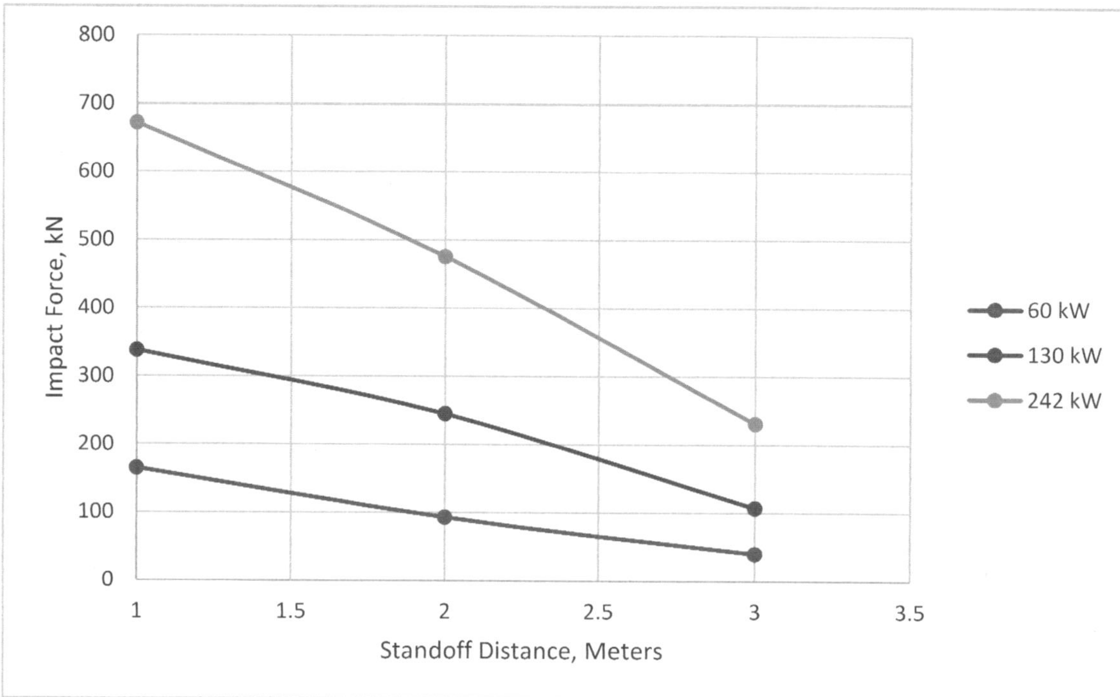

Impact Force by 276 MPa (40,000 psi) at Three Different Powers
Figure 3.

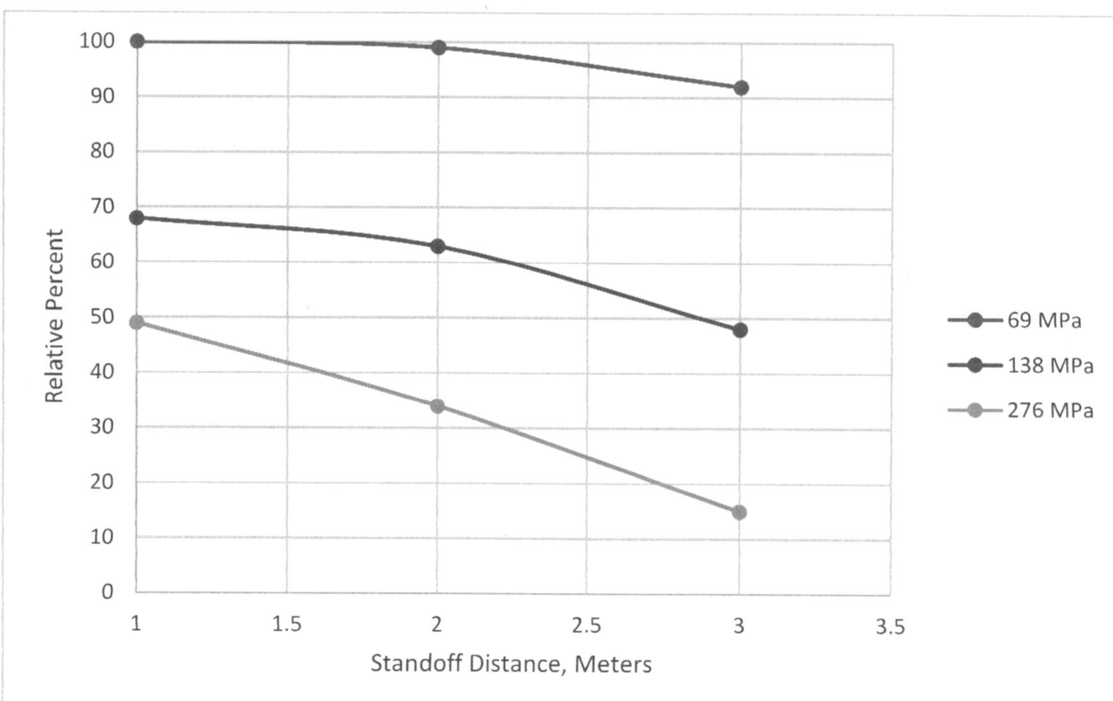

Average Relative Jet Impact Force by Pressure
Figure 4.

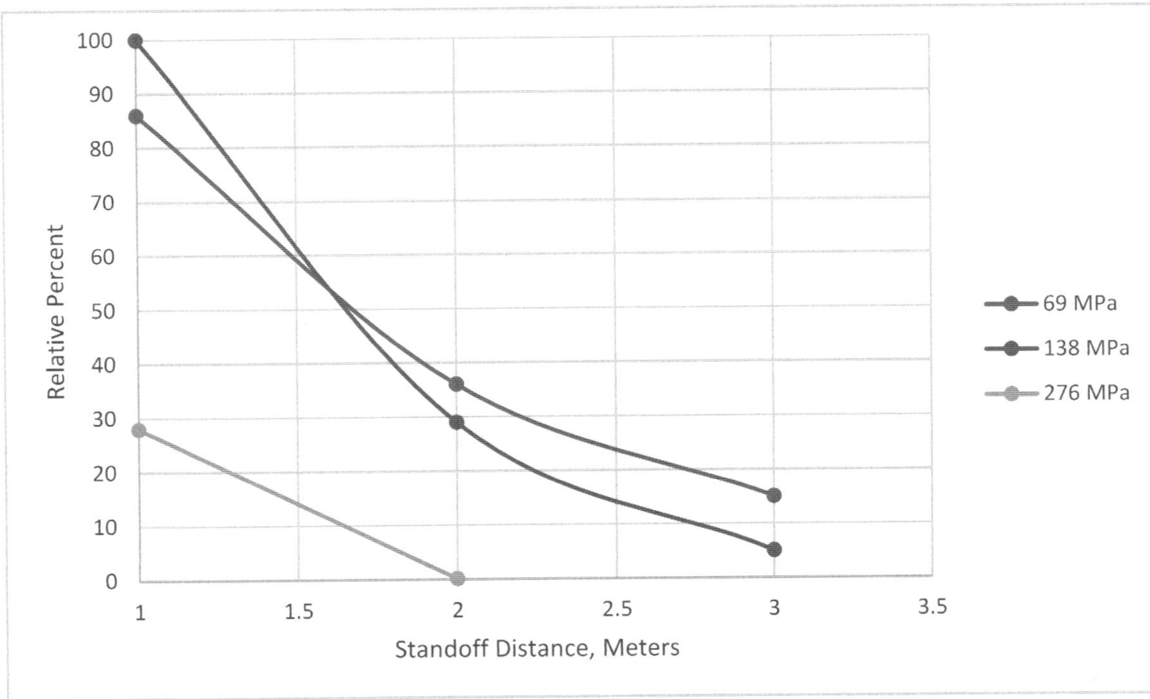

Average Relative Depth of Cut by Pressure
Figure 5.

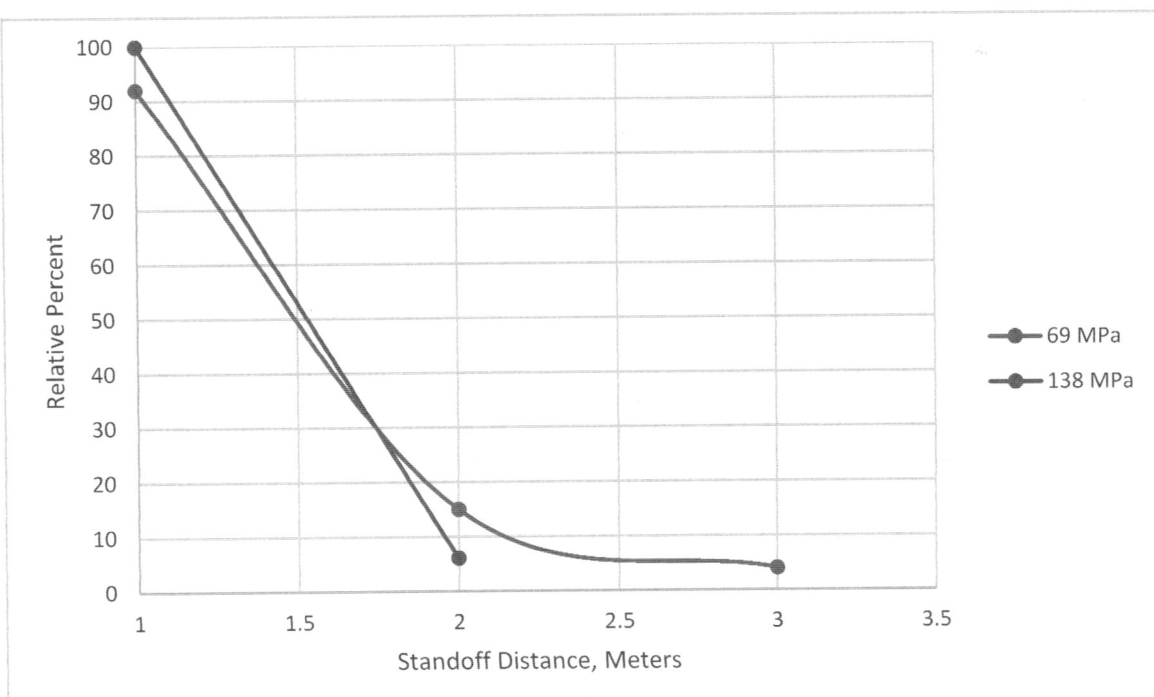

Relative Depth of Cut by Pressure at 60 kW
Figure 6.

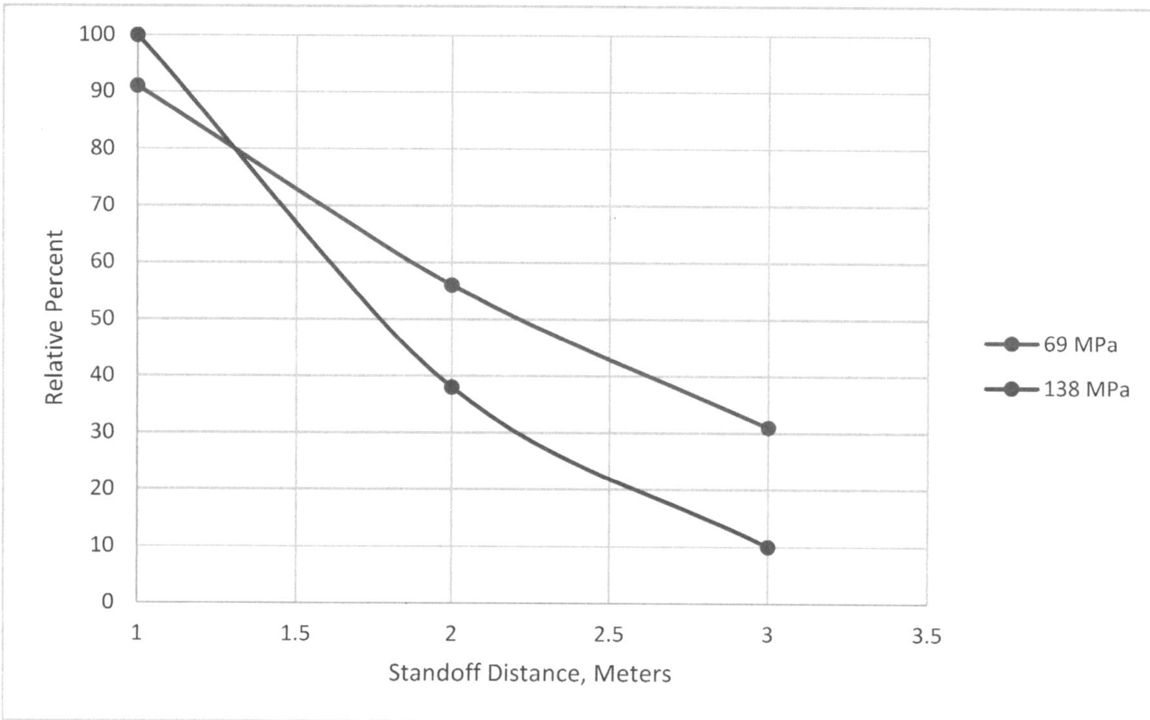

Relative Depth of Cut by Pressure at 242 kW
Figure 7.

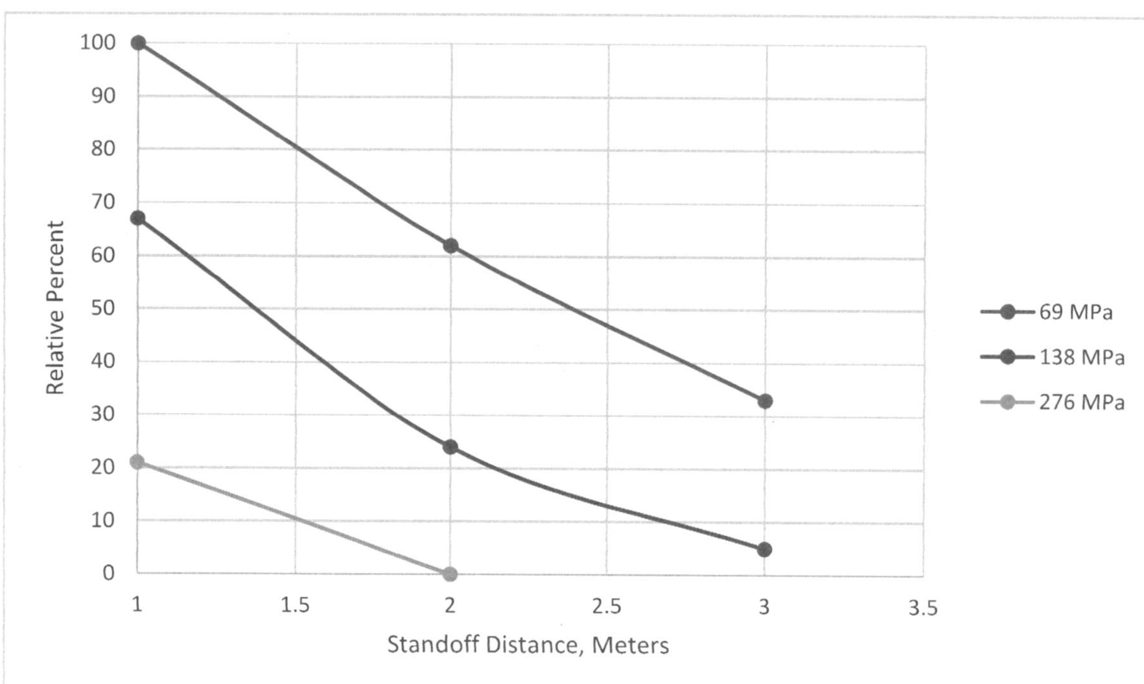

Impact Force Multiplied by Depth of Cut at 242 kW
Figure 8.

2019 WJTA Conference and Expo
November 11-13, 2019 • New Orleans, Louisiana

Paper

TECHNICAL COUNTERMEASURES RESEARCH ON DEEP SEA SHIPWRECK WATER CUTTING ENGINEERING

Shengxiong Xue, Zhengwen Chen, Qile Ren, Yongqiang Wang, Songlin Cao, Caihong Han, Di Zhang
Hefei General Machinery Research Institute, Anhui Province Key Laboratory of General Machine Composite Material Technology
Hefei, Anhui, China

ABSTRACT

The emergency disposal method for deep sea shipwreck which contains fuel and liquid hazardous chemicals is firstly to discharge the loaded liquid. This aim is to create a safe condition before salvage. As for this engineering, this paper introduced the first step technical method for shipwreck salvage engineering. That is to use water cutting equipment to cut a 1-2 square meter rectangular or circular hole for underwater ship outer hull. And then install special equipment in the inner ship hall for drilling, plugging and pumping liquid. For 6000 meter deep sea shipwreck water cutting engineering, the whole set of cutting equipment is proposed to submerged and cooperated with ROV robot. The overall conception and working aim of deep sea shipwreck water cutting equipment was analyzed. And types and parameters of pre-mixed abrasive water cutting was also analyzed. The key technology of submarine pump, sand mixing container and water cutting head were summarized. Series of test development and the key nodes were designed. The above technical countermeasures are the whole process implementation plan for this set of deep-sea shipwreck water cutting equipment.

1. ENGINEERING BACKGROUND

At present, the shipping industry in China and the world has developed greatly. As the density of ships increases rapidly, the risk of ship accidents is increasing. When a deep sea shipwreck accident occurs, in order to discharge and recycle dangerous chemicals from the shipwreck carrying or loading, it must be adopted emergency disposal such as opening, drilling, plugging and pumping liquid for shipwreck [1-2]. As for double hulls of shipwreck outer and inner hulls, breaking and opening operation of shipwreck outer hull is the first step of emergency disposal procedure, that is to create a safe precondition for other shipwreck salvage procedures.

Water cutting is a mature and advanced technology developed both domestic and overseas during nearly twenty years [3-4], it has many technical advantages. Because of its no selectivity for the cut material, water cutting processing is widely used in application. Water cutting is a cold processing and does not produce any thermal accumulation effect. It has characteristics of high cutting quality and cutting speed. Furthermore, it can be realized to remote control and program operation. So it is quite suitable for a large amount of cutting application.

The deep sea environment is quite harsh, and the underwater depth of the shipwreck is hundreds of meters or even thousands of meters. Therefore, the salvage engineering operation for deep sea shipwreck is quite difficult. It needs higher short time efficiency for shipwreck outer hole opening, and it needs breaking and opening equipment to meet higher adaptability requirements of deep sea operation conditions. Water cutting technology can use seawater as the working medium, it can be realized remote operation especially with the development of ROV (Remote Operated Vehicle) technology [5-6], and water cutting technology has become the preferred technology for deep sea shipwreck outer hall breaking and opening salvage. However, there is not any application report around the world about a whole set of equipment submerged into deep sea for shipwreck water cutting. In 2008, the Norway company has used water cutting technology to cut Russian Kursk submarine under the Barents Sea. But this water cutting equipment was still running above the seawater. This paper introduced Chinese technology countermeasures of 6000 meters deep sea shipwreck water cutting engineering which has already been implemented.

2. DESIGN & SCHEME RESEARCH

2.1 Overall Conception and Working Aim of Deep Sea Shipwreck Water Cutting Equipment

On the premise of adapting to deep sea environment, shipwreck water cutting equipment is to form a stable and reliable water jetting conditions. And it is to achieve functions of straight line cutting, large diameter circular cutting, large square hole cutting and so on. It is also to complete the ship outer hull forming hole cutting, ship hull stiffened plates and bars attachment cutting. Then the shipwreck is applied to other special equipments close to inner hull for hole drilling, plugging and oil pumping operations and so on.

To achieve the above goals, deep sea shipwreck water cutting application should face technical problems as follows.

1) It should be to produce high parameters of water jetting in deep sea for effectively cutting 15mm thick of ship steel hull.
2) Cutting area should be as large as possible and meet operation requirements of other special equipments for shipwreck inner hull drilling, plugging and liquid pumping.
3) Because the abrasive used for water cutting is limited carried, the once work time under water is about 30min.
4) It should be realized to remote water cutting operation in deep sea environment.
5) As for 6000 meters deep sea, it needs to overcome ultra high confining pressure negative impact on the submerged water jetting.
6) Deep sea environment adaptability of shipwreck water cutting equipment.

To deal with the above problems, the technical scheme of deep sea shipwreck water cutting engineering includes as follows. Considering the resistance influence on water jetting produced by 6000 meters ultra high confining pressure and water cutting equipment operation reliability easily ensured in low limit section of water cutting pressure, pre-mixed abrasive jetting technology which is increased to 100-140MPa from the current level of pressure 50-60MPa. It can avoid complex structure and deep sea operation inadaptability of common mixed abrasive jetting technology. To meet requirements of forming large area cutting operation and remote control automatic water cutting, the whole set of water cutting equipment is cooperated with and powered by underwater ROV robot. Water cutting head is equipped with special adsorption guide component which is designed and installed on ship outer hull. It is to achieve self water cutting two-dimensional along the adsorption guide which is remote controlled by ROV, and the head's travel speed can be adjusted. In order to achieve higher cutting efficiency under water, the whole set of deep sea shipwreck water cutting equipment is optimized to extremely match the parameters of pre-mixed abrasive mixing vessel, ultra high pressure pump flow, water cutting head, water cutting time, water cutting speed and water cutting volume. A series of tests such like ground test, underwater test, pressure resistance simulation test and deep sea simulation test [7] are carried to explore the adaptability of the whole water cutting equipment. On the basis of the above technical scheme, the working aim of shipwreck water cutting equipment is to cutting a 1-2m^2 rectangular or circular large hole in the outer of the shipwreck hall.

2.2 Scheme of Deep Sea Shipwreck Water Cutting Equipment

The key technologies of deep sea shipwreck water cutting equipment: to develop ultra high pressure pre-mixed abrasive water cutting device, make the water cutting head travelling and cutting along the adsorption guide component has been installed on ship outer hull, integrate the whole set of equipment and submerge into the deep sea, and effectively cooperate with ROV and complete ship outer hull water cutting. The whole set of deep sea shipwreck water cutting equipment includes as follows.

1) Ultra high pressure sea water submarine pump unit shown in **Figure 1** is considered a generating device for ultra high pressure water jetting. The whole sea water submarine pump unit is an integrated frame unit and submerged into deep sea. In the deep sea environment, this main submarine pump of unit is a reciprocating pump to pressurize seawater and it is driven by a hydraulic motor which is powered by an underwater ROV robot.

Figure 1. Ultra high pressure sea water submarine pump unit

2) The pre-mixed abrasive mixing system uses 80 mesh garnet as abrasives. It is composed of abrasive mixing switch valve and two abrasive mixing vessels which are one work and one backup. This mixing system is used for storage abrasive and bearing 100-140MPa of ultra high pressure. And then it forms pre-mixed abrasive water jetting through abrasive mixing switch valve and abrasive nozzle.

3) Ultra high pressure water cutting executive device shown in **Figure 2** is the executive mechanism of deep sea shipwreck water cutting equipment. It contains water cutting head, adsorption guide component, driving mechanism, hose lines, etc. It matches and realizes fast connection with ROV interface. The water cutting executive device is wholly independent, which is suitable for engineering requirements of cutting shape and size. It is self stable adhesive to ship outer hall and combined with water cutting head. The driving mechanism controls water cutting head travelling along adsorption guide and keep the best target distance of water cutting. The water cutting head is remote controlled and to achieve two-dimensional direction operation. Abrasive mixing switch valve is also remote controlled to achieve switching function for abrasive water jetting.

Figure 2. Ultra high pressure water cutting device cooperated with ROV

4) Ultra high pressure water cutting head is fixed travel along adsorption guide of water cutting executive device. The abrasive mixing is straightly formed before abrasive nozzle, and pre-mixed abrasive jetting is sprayed by abrasive nozzle.

3. PARMETERS & DEVELOPMENT RESEARCH

3.1 Type and Principle of Deep Sea Shipwreck Pre-mixed Abrasive Water Cutting

Abrasive water cutting technology involves a certain amount of abrasive particles into the pure water jetting. After abrasive and water mixing, the abrasive particles can be used to impact material with high speed of water jetting. It is greatly improving the ability of water jetting penetrating material.

A common mixed abrasive water cutting technology [8-9] shown in **Figure 3** uses water nozzle to spray water jetting and forms vacuum to absorb abrasive into the mixing chamber, and finally forms liquid and solid mixing jetting to spray from the abrasive nozzle. Because the acceleration time of absorbed abrasive is short and the water jetting after sprayed is quickly put into basic stage and dissipation stage of water jetting, it caused that the kinetic energy of accelerated abrasive is limited. So the water jetting pressure of common mixed abrasive water cutting materials must be reached 200-400MPa [10].

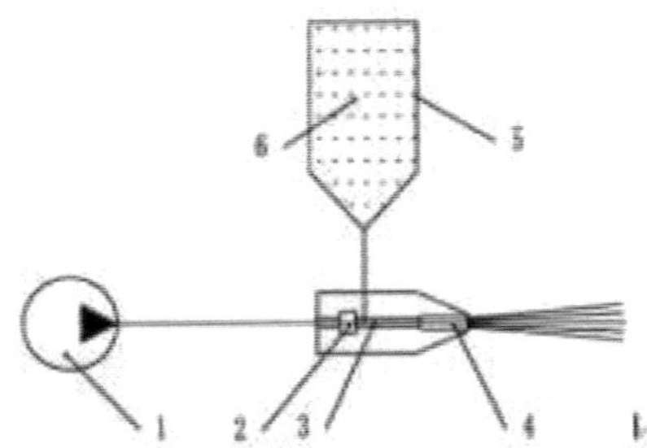

Figure 3. Principle of common mixed abrasive water cutting
1.Pump 2.Water nozzle 3.Mixing chamber
4.Abrasive nozzle 5.Abrasive tank 6.Abrasive

The shipwreck is in the extremely high confining pressure which is environment pressure. The confining pressure of deep sea shipwreck has resistance and damage to the formation and condensation of water cutting jetting. The higher the confining pressure is, the worse efficiency of water cutting jetting is. Deep sea shipwreck water cutting technology uses pre-mixed abrasive water cutting technology shown in **Figure 4**. That is, a part of ultra high pressure water (sea water) is put into abrasive mixing vessel and other ultra high pressure water is put into abrasive nozzle, abrasive and water is mixed and accelerated before abrasive nozzle, then abrasive water jetting of liquid and solid two phase jetting is sprayed from abrasive nozzle. As the abrasive is fully mixed in the water jetting and prolonged the acceleration time, the working pressure of pre-mixed abrasive water cutting is only 50-60MPa for the same cutting material. Because the pressure of pre-mix abrasive water cutting is reduced to the low level of ultra high pressure which is 100MPa of starting point, the reliability of high pressure equipment has been greatly enhanced and the working pressure margin to overcome confining pressure of deep sea has also been improved. It is no doubt that the pre-mixed abrasive water cutting is the best technical scheme of deep sea shipwreck breaking and opening salvage.

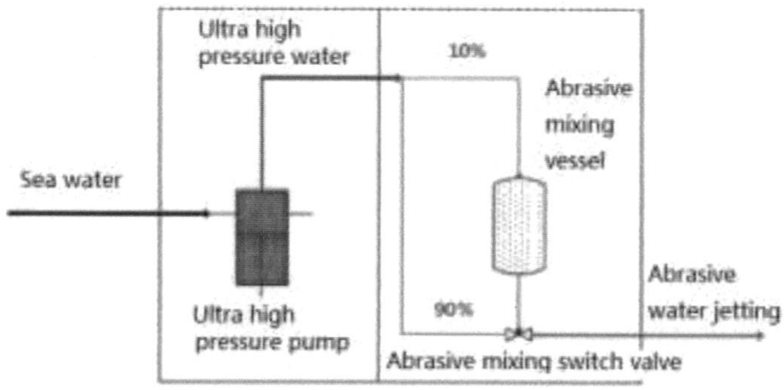

Figure 4. Principle of pre-mixed abrasive water cutting for deep sea shipwreck salvage

3.2 Working Parameters of Pre-mixed Abrasive Water Cutting in Deep Sea

It must be firstly faced harsh environment condition of 60MPa confining pressure about 6000 meters deep sea shipwreck water cutting engineering. In the domestic and foreign research and application, whether it is deep surface oil drilling or coal seam excavation, the confining pressures of these applications are both less than 1000 meters and these equipments are basically working on the ground. The real application with ultra high confining pressure environment which is under water is not common, but only there is a little application of the constraint conditions such as oil casing pressure. The direct and effective means of overcoming the confining pressure is to increase the water jetting working pressure. The author has carried out the ground experiment of pre-mixed abrasive water cutting and the experimental data are shown in **Table 1**.

Table1. Ground experiment data of pre-mixed water cutting

Parameters	Experiment situation					
Pressure(MPa)	200	140	100	140	100	60
Flow(L/min)	10	17	27	10	15	15
Power of pump(kW)	55			30	30	22
Water cutting speed of 10mm steel plates(mm/min)	800			600	400	300
Water cutting speed of 15mm steel plates(mm/min)	600			400	300	240

It can be seen from experiment that the better performance for water cutting 10-15mm steel plate is under 60MPa of pressure and 15L/min of flow. With the further incensement of pressure and flow, the efficiency of steel plate cutting is higher. When the pressure increases to 140-200MPa, the efficiency of steel plate water cutting is limited.

According to the experiment results of pre-mixed abrasive water cutting on the ground and the effect of ultra high confining pressure, the parameters of deep sea shipwreck water cutting engineering are 100-140MPa of pressure and 10-15L/min of flow. The pressure is distributed as follows. 1) The normal working pressure of pre-mixing abrasive water cutting is 60MPa. 2) The remote pressure losses and the seabed working pressure margin taken into account is 20MPa. 3) The pressure to overcome ultra high confining pressure of 6000 meters deep sea is 60MPa. 4) Because the above two pressures are evaluated relatively conservative, such as 6000 meters is limit depth, 100MPa of pressure and 15L/min of flow with equal power is used as test conditions.

Resistance force caused by ultra high confining pressure is still in qualitative and not quantitative research. Although ultra high confining pressure which is static pressure can greatly improve the inlet pressure of pump unit and cannot improve the outlet pressure of pump unit, it is only benefit to save the pump unit running energy because pressure of water jetting is produced by nozzle. In a word, confining pressure of seabed water jetting is a resistance force.

3.3 Key Technology of Submarine Pump

In order to obtain a stable and reliable ultra high pressure water cutting condition, the structure of the submarine pump is designed a three-plunger ultra high pressure pump shown in **Figure 5**. The pump structure is self balanced. The main body of this pump is composed of the power end and the liquid end. The submarine pump is submerged into the seawater environment, inside and outside of the pump is loaded the ultra high confining pressure. The material of the exposed parts uses a special high strength double phase stainless steel material to ensure the bearing confining pressure. The key components such as dynamic seals etc. are corrosion resistant materials and composites. And they use sea water to lubrication. In order to apply for deep sea environment, the driving unit of the submarine pump is a hydraulic motor, its hydraulic characteristics are highly matched with the liquid characteristics of the submarine pump and 300rpm of pump rotary speed.

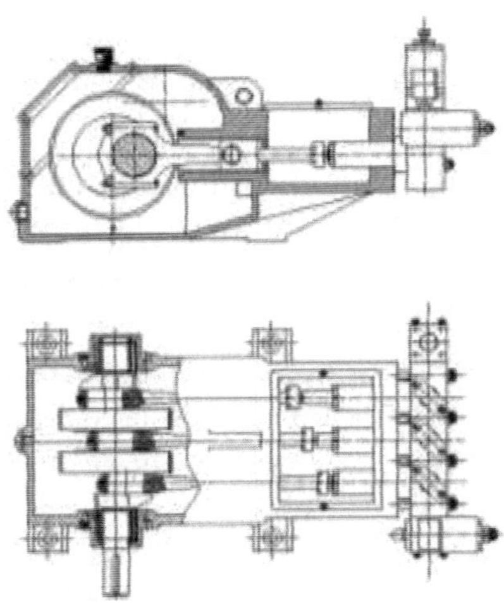

Figure 5. Structure of submarine pump

A notable feature of the submarine pump is that all rotating parts of power end are not rolling bearings, but sliding bearings and axial balancing systems. Because rolling bearings are easily to damage or even seized without oil lubrication in the seawater environment.

3.4 Key Technology of Abrasive Mixing Vessel

As ultra high pressure vessel, abrasive mixing vessel is pressurized 140MPa. It is difficult to improve the volume under this pressure. The whole set of deep sea shipwreck water cutting equipment uses two abrasive mixing vessels, every vessel is designed 25L of volume and 100kg of storage. After 30min water cutting operation using one abrasive mixing vessel, another spare abrasive mixing vessel can be switched to use according to the needs.

The abrasive mixing vessel is bolted to the head, and it is inclined reliably and fixed. The ultra high pressure water of deep sea shipwreck water cutting is entered from the upper end of the vessel, and the abrasive is mixed and absorbed from the bottom of the vessel. The abrasive mixing switch valve is installed in the bottom of this vessel, and the consumption of the pre-mixed abrasive is adjusted about 3kg/min. The Abrasive mixing vessel is resistant to seawater corrosion. The ultra high pressure water is entered into chamber and is mixed with dry abrasive, then it is realized realize uniform and stable supply of abrasive mixing through switch valve.

3.5 Key Technology of Water Cutting Head

After mixing the water with abrasive, the ultra high pressure abrasive water jetting is sprayed through the water cutting head. As the final executive component, it is not only subjected to seawater corrosion but also it has good abrasion resistance to the high speed abrasive water jetting. The design of water cutting head is matched with the parameters of pre-mixed abrasive water jetting. The bore of the abrasive nozzle should have a considerable ratio of length to diameter, which is conducive to the condensation and straight line of abrasive water jetting.

3.6 Series of Tests Development and Its Key Points

The whole set of deep sea shipwreck water cutting equipment and application cannot be separated from engineering test. Series of tests will play a guiding role in improving equipment development, forming commercialization and standard application. The series of tests include as follow.

1) Equipment performance parameters tests. It is carried out according to the test method specified by GB/T 26163 ultra high pressure water cutting machine and JB/T 6909 ultra high pressure pump. Through the performance parameters experiments of the submarine pump, abrasive mixing vessel, water cutting head, other components and the whole set of equipment, the design parameters are confirmed to achieve the expected design target.

2) Underwater test. It is carried out with simulation device of using deep sea shipwreck salvage. According to the emergency salvage procedure, the effect of ultra high pressure water cutting experiment in conventional tanks is gradually tested, and the operation reliability of the whole equipment under submerged conditions is confirmed by this test.

3) Pressure resistance simulation test. It is adopted high pressure container which is used for simulation 6000 meters deep sea confining pressure environment. This experiment is to put the components into ultra high confining pressure, the whole set of equipment pressure resistance test is depending on the overall terms and conditions. This experiment includes static pressure resistance test, dynamic pressure resistance test, performance test, remote monitoring operation of equipment components. The performance of the equipment under ultra high confining pressure condition is confirmed by this test.

4) Deep sea test and application. On the premise of meeting the requirements of above tests, deep sea shipwreck water cutting equipment is submerged into deep sea by the Diving salvage ship and coordinated with ROV shown in **Figure 6**. To achieve the working aim of application, the deep sea experiment is to realize the complete underwater operation such like the connection of submarine, adsorption guide laying and installation, achieve underwater power, underwater start-up and remote control for 30-50 meters, deep sea water cutting operation and so on.

a) The Diving salvage engineering ship

b) ROV robot for deep sea shipwreck salvage

Figure 6. Salvage ship and ROV

4. CONCLUSION

Pre-mixed abrasive water cutting technology of 100-140MPa is suitable for 6000 meters deep sea shipwreck water cutting engineering. Proof basis and technical scheme are convincing and feasible. Ultra high confining pressure, ultra high pressure abrasive water jetting in deep sea, abrasive mixing and control, water cutting operation are the key technical countermeasures for deep sea shipwreck water cutting engineering. The whole set of deep sea shipwreck water cutting equipment, make ultra high pressure submarine pump and ultra high pressure water cutting device coordinated with ROV as the core components of shipwreck salvage, can completely adapt to deep sea operation requirements. Series of tests including four tests are important parts of equipment development, and they are an indispensable progressive part. It still needs to be continually modified and adjusted for above technical countermeasures in the implementation processing. Especially it is necessary to refine the design and test for underwater less action frequency and even non action.

5. ACKNOWLEDGMENTS

The work of this paper is supported by the National Key Research Support Project of China (2017YFC0307003).

REFERENCES

[1] Yinliang Zhang, Zhigang Li, Yinbin Feng, etc. "Study on Ways to Deal With Oil Spill Accident of Deepwater Ship by Combination of ROV and Diver," EVIRONMENT ENGINEERING, 2013, 31(S1): pp54-57.

[2] Weiting Ning, Jiahai Zhou, Yunfei Zou, etc. "Study on Ways of Underwater Drilling and Oil Pumping of Oil Tank on Sunken Shipwreck," SHIP ENGINEERING, 2015, 37(10): pp106-110.

[3] Shengxiong Xue, Zhengwen Chen, Qile Ren, etc. "Research on Application of 500MPa Water Cutting Composite Material," Proceedings of the ASME 2017 Fluids Engineering Division Summer Meeting, FEDSM2017, July 31-August 3, 2017, Waikoloa, Hawai'i, USA

[4] Shengxiong Xue, Zhengwen Chen, Bo Chen, etc. "Introduction and Analysis of the Ultrahigh Pressure Water Jet Cutting Multifunctional Application," FLUID MACHINERY, 2017, 45(11): pp17-21.

[5] Shitao Li. "Application Research of ROV Technology to Deep Water Engineering," SHIPBUIDING OF CHINA, 2009, 50(11): pp222-228.

[6] Yang Dong. "The Research of Black Box Dedicated Grab Tool In Deep Sea," Maritime Affairs University of Dalian, 2016.

[7] Qiang Zhang, Leili Zhang, Mingjun Zhang. "Experiment devices for simulating a deep-sea environment and dynamic pressure control technology," Journal of Harbin Engineering University, 2016, 37(11): pp1565-1572.

[8] Shengxiong Xue, etc. "High Pressure Water Jetting Technology Application," China Hefei University of Technology Press, 2006:pp485-498.

[9] Xinping Long, Qi Liu, Xiaofeng Ruan, etc. "Numerical Analysis of Internal Flow of Abrasive Entrained Waterjet nozzle," Journal of Drainage and Irrigation Machinery Engineering, 2016, 36(8): pp686-692.

[10] Shengxiong Xue, Junkun Meng, Qile Ren, etc. "Research on Working Condition and Cutting Mechanism of Water Cutting Equipment for Large Scale Curved Composite Component," FLUID MACHINERY, 2016, pp44(12):25-29.

2019 WJTA Conference and Expo
November 11-13, 2019 • New Orleans, Louisiana

Paper

ABRASIVE SURFACE TREATMENT WITH LIQUID CO_2

E. Uhlmann, P. John, C. Männel
Technische Universität Berlin
Berlin, Germany

ABSTRACT

The main advantages of cutting with liquid jets are the flexibility and consistent sharpness of the tool, which allows the machining of a variety of materials and complex shapes. Unfortunately, the humidification of the components can be a problem for certain applications and inhibits the spread of jet technology. Besides, the dry and residue-free cutting of materials is an important topic of today's research in manufacturing engineering. Due to these advantages, high-pressure liquid CO_2 jet cutting has the potential to open new fields of applications in which water jet cutting is not suitable. The liquid CO_2 jet with a pressure of up to 300 MPa can be used to machine various materials and functional surfaces before it expands to gas and atmospheric pressure. However, the transition from liquid to gaseous phase implicates density differences, which change the cutting performance. As a result, the knowledge about waterjets cannot be adapted to CO_2 jets and further investigations are necessary.

At a prototypical test stand for high-pressure liquid CO_2 jets, a feed line with abrasives was added. Technological investigations concerning the surface properties with high-pressure liquid CO_2 and water jets were performed with and without different abrasives. The tests were carried out on parts of various metals and technical plastics. The influence of the fluid on jet properties as well as the attained workpiece surfaces and kerfs produced by the different jets were investigated.

The experiments indicate that the performance of the CO_2 jet as well as of the waterjet mainly depends on pressure, focus tube and nozzle diameter but show different behavior. Especially the impact of the working distance and the jet temperature will be discussed. The investigations reveal that high-pressure liquid CO_2 jets have a high potential in the field of dry and residue-free cutting and dry surface treatment, like peening, of metals, technical plastics and CFRP.

Organized and Sponsored by WJTA®

1 INTRODUCTION

Jet machining has established itself for years as an efficient and progressive manufacturing process and has been enhanced continuously. The manufacturing process produces kinetic energy by means of a continuous jet of liquid, usually water, which can be used to implement various manufacturing options and new fields of applications. On the one hand, the well-known omnidirectional cutting effect can be used to process a wide range of materials, which are subjected to both low mechanical and low thermal loads during the low-wear manufacturing process. On the other hand, the resulting negative pressure in the abrasive tube and the special properties of water can also accelerate materials in powder form, so-called blasting abrasives. This can either increase the cutting performance, as in abrasive water jet cutting [1, 2], or the material properties of the surface can be changed, as in shot peening [3].
However, for some applications water will never be the first choice of jet medium or machining technology due to its additional needed machine periphery and process steps, such as the disposal of water or cleaning and drying of the workpieces after machining [2]. In contrast, comparable blasting technologies with air or blast wheels for surface treatment cannot accelerate the abrasive sufficiently or are not suitable for certain geometries [3, 4].
Debunking these limiting factors of conventional processes, while simultaneously using the advantages of jet machining, the liquid carbon dioxide (CO_2) jet is an alternative acceleration medium with high potential for new applications. Due to the complete sublimation of the jet, high-pressure jet cutting with liquid CO_2 is a dry and residue-free process. It was first investigated in a feasibility study by DUNSKY and HASHISH [5], proving the realisability of the process under atmospheric conditions and showing similarities to water jet cutting as a residue-free cutting process. Based on these results, BILZ [6] designed a prototype system at the Production Technology Centre Berlin (PTC) and continued with detailed analytical and experimental investigations. Force impulse measurements and the evaluation of kerf characteristics on plastic specimens were conducted to show the industrial potential of the high-pressure jet cutting process. Originating from a joint research project with the PTC, ENGELMEIER [7] carried on investigations in order to analyze pressures and temperatures in the process and their influence on jet deformation and decay. In previous work, a general suitability of the process for a dry and residue-free cutting of metal materials was proven [8]. Investigations on jet velocity, jet impulse force and kerf geometry on aluminum specimen ($AlMg_3$) using jet pressures up to 300 MPa led to knowledge about main differences between water jet cutting and jetting with liquid CO_2. Continuing, pressure, nozzle diameter, but also the working distance, secondary material damage and the potential of accelerating abrasives was examined [9, 10, 11]. This paper will investigate the depth of cut for aluminum AlMg3 depending on various influencing factors. Not only obvious process parameters like pressure and nozzle diameter, but also the working distance, the jet temperature and the adding of different abrasives will be examined.

2 TEST STAND AND MEASUREMENT SETUP

2.1 Test stand for liquid CO_2 jets

By using the prototype system [6], functional correlations between significant setting parameters and results were developed analytically and experimentally in order to analyze the cutting properties of the CO_2 jet. Following, the test stand and the measuring principle to analyze these quantities are described.

The empirical investigations on the plain water jet cutting technology were performed with the system Jet Max HRX 160L from MAXIMATOR JET GMBH, Schweinfurt, Germany, with a high-pressure pump HPS 6045 which realizes jet pressures up to p_0 = 600 MPa.

The liquid CO_2 cutting jet system is divided into three functional modules: Climatic chamber, high-pressure pump and cutting chamber. The liquid CO_2 is supplied from a riser pipe bottle inside the chamber via high-pressure hoses to the suction side of the high-pressure pump. According to the temperature inside the climatic chamber, the supply pressure is regulated up to p_V = 9 MPa at T = 45 to 50 °C. Within the pump, liquid carbon dioxide is gradually compressed up to 300 MPa and pumped to the pulsation damper and subsequently to the closed cutting head in the cutting chamber. The high-pressure pump is a Steamline1 of INGERSOLL-RAND, Swords, Ireland, with a maximum pressure of p_0 = 345 MPa and a maximum flow rate of Q = 3.8 l/min. The cutting head is pneumatically actuated and opened, allowing the high-pressure fluid to exit the nozzle. The cutting head, Active Autoline II from KMT GMBH, Bad Nauheim, Germany, attached to a gantry robot is moved according to the chosen direction and at the selected feed speed. This experimental setup allows the machining of parts with high-pressure CO_2 jets of different length and power, depending on the influencing variables (Figure 1a-1f).

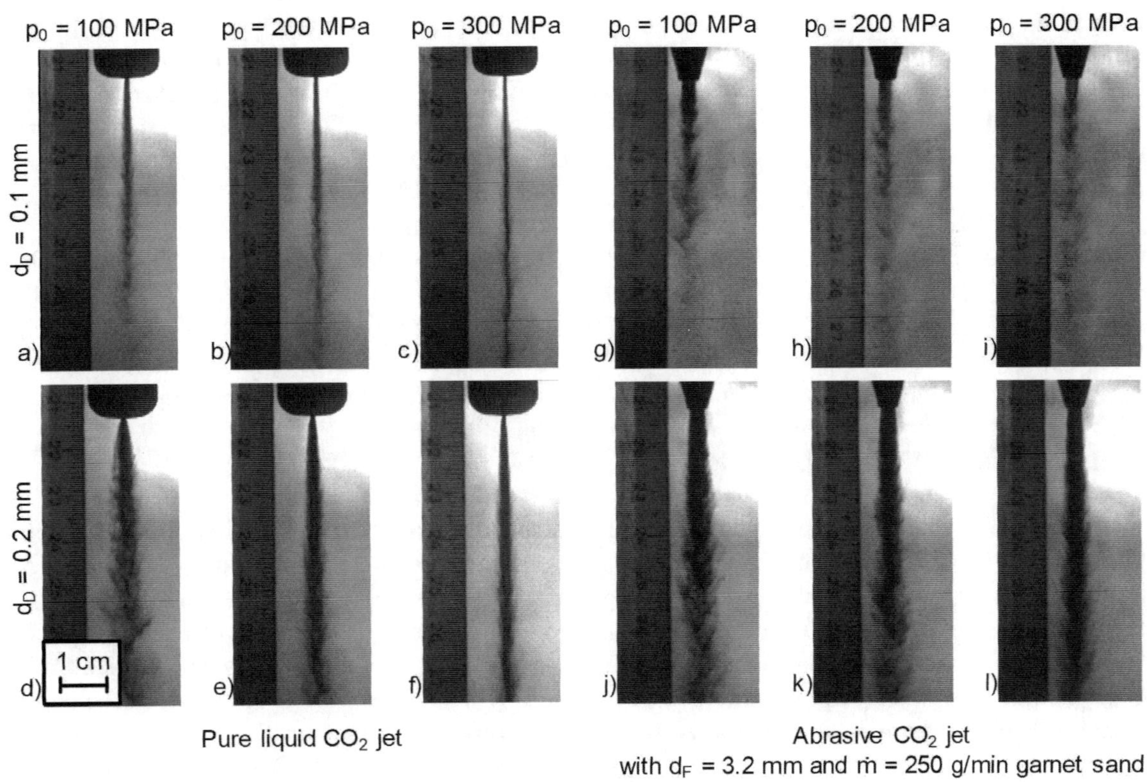

Figure 1. High-pressure liquid CO_2 jets [11].

In order to increase the removal rate of the jet, abrasives such as garnet sand, quartz sand or corundum are used in abrasive waterjet cutting. Thus, hard materials such as steel, stone or glass

can also be processed economically. The proof of concept and the industrial potential of abrasives added to the liquid CO_2 jet for cutting has already been demonstrated with a prototype out containing many waterjet components [10]. To investigate this, water abrasive injection jet cutting was chosen as an overall concept. It is established in the industry and their sophisticated components for which spare parts and technical service are already available. This offers the simultaneous advantage that existing systems do not have to be modified at all or only slightly. Thus, for the dosing of the abrasive a commercial abrasive dosing unit and feed as well as further required parts with already existing technology were selected. The experimental investigations were amended by the FEEDLINE V and by the focusing tubes HYPERTUBE 54x947x50-21-A1, both of the KMT GMBH, Bad Nauheim, Germany. The abrasive feed system is able to realize a mass flow up to $\dot{m} = 1000$ g/min.
The major challenge hereby is to realize a vacuum in the abrasive hose and getting the abrasive into the mixing chamber or afterwards into the jet. Therefore, various parameters like the jet pressure p_0, the nozzle diameter d_D and the focusing tube inner diameter d_F have to be verified. The different focusing tube inner diameters d_F were adjusted within the investigations through eroding.
In order not to lose the process advantage of a residue-free process, various additives were tried out. In this paper, Garnet sand with Mesh 140, grain size $d_G = 0.14$ mm, from BARTON INTERNATIONAL PLC, Glen Falls, USA, was the choice for machining metals. In addition, to maintain the advantage of residue free machining, CO_2 crystals have been added to the jet instead of garnet sand. The CO_2 crystals had to be smaller than $d_G = 0.6$ mm in order not to block the focus tube and the mixing chamber. Therefore, two different possibilities were carried out. In a first feasibility test, conventional CO_2 pellets, from PRAXAIR DEUTSCHLAND GMBH, Düsseldorf, Germany, have been chopped and sieved by hand. In the second approach, a supplementary machine system, i³ MicroClean DX from COLD JET GMBH, Weinsheim, Germany, was used for shredding a CO_2-cube and accelerating the particles into the cutting head.
However, different material densities and powder sizes cause a different mass flow. Therefore, the abrasive feed was calibrated for each powder. With a precision balance of the type PLS 1200-3A from KERN & SOHN GMBH, Balingen, Germany, the correct potentiometer settings were found. Finally, the experimental setup allowed the machining of parts with high-pressure abrasive CO_2 jets of different length and power, depending on the influencing variables (Figure 1g-1l).

2.2 Measurement of kerf characteristics

To analyze the jet quality, the potential of creating kerfs is a valid method [6]. Therefore, the depth of cut k_T, kerf width k_B, middle kerf width k_{BM} and kerf shape k_F were investigated. As a result of the special cutting behaviour of plastics, the depth of the kerfs is divided into the parameters depth of cut k_T and penetration depth k_{ET}. The penetration depth k_{ET} is defined as the maximum depth at which an influence of the jet is still visible, whereas the depth of cut k_T only includes the area in which the material was visibly separated. The middle kerf width was measured at half of the depth of cut k_T. The kerf shape k_F is a qualitative value and compares the kerf profile along the abscissa axis by cutting vertically through the kerf (Figure 2). For these investigations the specimen of rolled sheet with a thickness of 5 mm consisting of the aluminum alloy $AlMg_3$ were processed with different parameters. The material was chosen due to the characteristic properties of metal, but with low hardness. Previous results of the jet impulse forces F_S [8] as well as preliminary tests [10, 11] led to the parameter field shown in Table 1. To provide statistically firm results, kerfs with a length of $k_L = 10$ mm were processed for each parameter variation and measured at three different locations. The measurements on the specimen were realized by using the chromatic white light interferometer FRT MicroProf and the chromatic sensor FRT CWL from FRIES RESEARCH & TECHNOLOGY GMBH, Bergisch Gladbach, Germany. To evaluate the data the software FRT Mark III from the same company was used. The second measurement system was the digital microscope VHX-5000 from KEYENCE DEUTSCHLAND GMBH, Neu-Isenburg, Germany.

Table 1. Parameters to determine the jet quality.

Explanatory variable	**Abb.**	**Input Values**	**Dimension**
Jet fluid	-	CO_2 / CO_2 + abrasive	-
Cutting head	-	Pure Jet / Abrasive jet	-
Jet pressure	p_0	100 / 200 / 300	MPa
Supply pressure	p_V	9	MPa
Nozzle diameter	d_D	0.10 / 0.15/ 0.20 / 0.25	mm
Working distance	a_W	5 / 10 / 15 / 20 / 25 / 30	mm
Jet feed speed	v_f	0 / 30 / 60 / 120	mm/min
Abrasive mass flow	$\dot{m}$	200 / 300 / 400 / 588	g/min
Abrasive material	-	Garnet / CO_2-microcrystals	-

a) 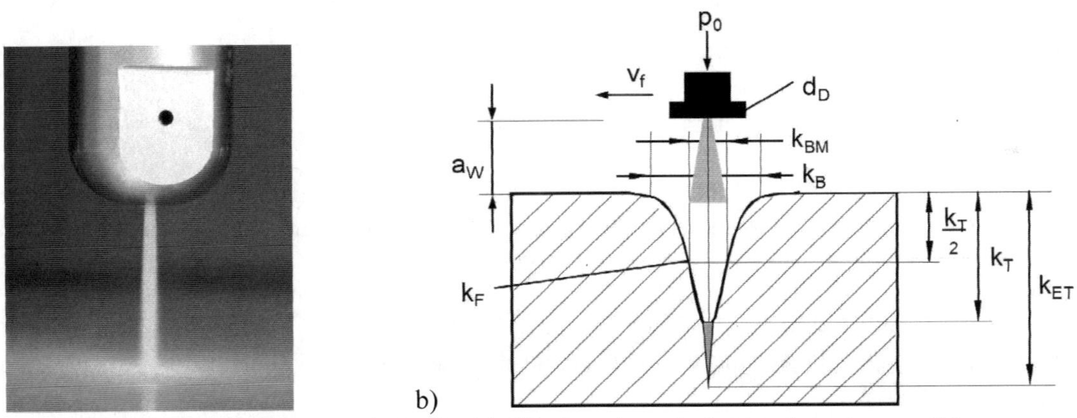 b)

Figure 2. Characteristic liquid CO_2 jet; a) optical; b) values to quantify.

2.2 Measurement of jet temperature characteristics

Considering the thermodynamic properties of CO_2 it is obvious, that the temperature is an important factor for the cutting properties of a CO_2 jet. The jet temperature T_S is an indicator of the phase transition of the CO_2 as well as its density. To analyze the sequence of the phase transition of the CO_2, the working distance is the most important parameter to be varied. Another aspect is the temperature development during the cutting process on the specimen. An example is thermoplastics, which have, compared to other materials, a low melting temperature and tend to become brittle at very low temperatures. In order to investigate the temperature development during the cutting process, the temperature curve during a jet time t_S has to be recorded. The tests are carried out without a specimen by jetting directly onto a thermocouple on a grid table. The first measuring point is located directly at the nozzle, the others along the jet to simulate certain working distances a_W. The parameters are shown in Table 2.

Table 2. Parameters to determine jet temperature.

Explanatory variable	Abbr.	Input values	Dimension
Jet fluid	-	CO_2	-
Cutting head	-	Pure Jet / Abrasive jet	-
Jet pressure	p_0	100 / 200 / 300	MPa
Supply pressure	p_V	9	MPa
Nozzle diameter	d_D	0.10 / 0.15 / 0.20	mm
Working distance	a_W	0 / 5 / 10 / 15	mm
Jetting time	t_S	5	s

The temperature is measured by a thermocouple type T from TC DIRECT, Moenchengladbach, Germany, which has a measuring range from T_S = -75 to 250 °C. It consists of two conductive wires of different materials. Type T thermocouples are made of copper and copper-nickel. Due to the thermoelectric effect, temperature changes generate a voltage in the microvolt range. This voltage is automatically converted into an absolute temperature by the connected measuring device and can be read directly. The thermometer TM-914C from LUTRON ELECTRONIC ENTERPRISE CO., Taipei, Taiwan, serves as the measuring instrument.

3 RESULTS AND DISCUSSION

3.1 Kerf characteristics for metals

The investigations were divided into jet tests with and without abrasives in order to compare them with each other. The pure jet tests were then again divided into normal jet tests and jet tests with the new abrasive jet head.

The pure jetting with the abrasive cutting head obtained no measurable results, but a cleaning of the surface was detected. This indicates that the distance to the workpiece is too high and the mixing chamber supports the phase transition, so that the density of the fluid is not sufficient to cut aluminium AlMg3.

For liquid jet cutting without abrasive cutting head, a jet pressure of p_0 = 300 MPa resulted in the deepest kerfs, which could be determined for all nozzle sizes. The maximum depth of cut with k_t = 335 µm was achieved at jet pressure p_0 = 300 MPa, nozzle diameter d_D = 0.15 mm and a jet distance of a_W = 4 mm (Fig. 3). The resulting kerf shape can be described as V-shaped. There is no flat kerf base. A dependence of the depth of cut on the nozzle diameter was also determined. The increase of the nozzle diameter from d_D = 0.10 to 0.15 mm causes an increase in the depth of cut k_T by a factor of 3.29. With a nozzle diameter of d_D = 0.10 mm, the volume flow is not yet sufficiently large to be able to achieve optimum removal through the described process. If the nozzle diameters are higher than d_D = 0.15 mm, the necessary volume flow seems to be too large for the machining system and the optimum was exceeded. Furthermore, it could be determined that the depth of cut increases to a certain optimum for every nozzle and afterwards decreases again with an increasing distance to the specimen. The kerf width k_B, not illustrated, increases with increasing distance to the specimen until no more kerf is cut. By increasing the distance from a_W = 2 to 30 mm, a widening of the kerf by factor 1.8 is possible. The increase is explained by the fact that the CO_2 partially passes downstream into the gaseous phase and loses depth impact but has a stronger jet decay.

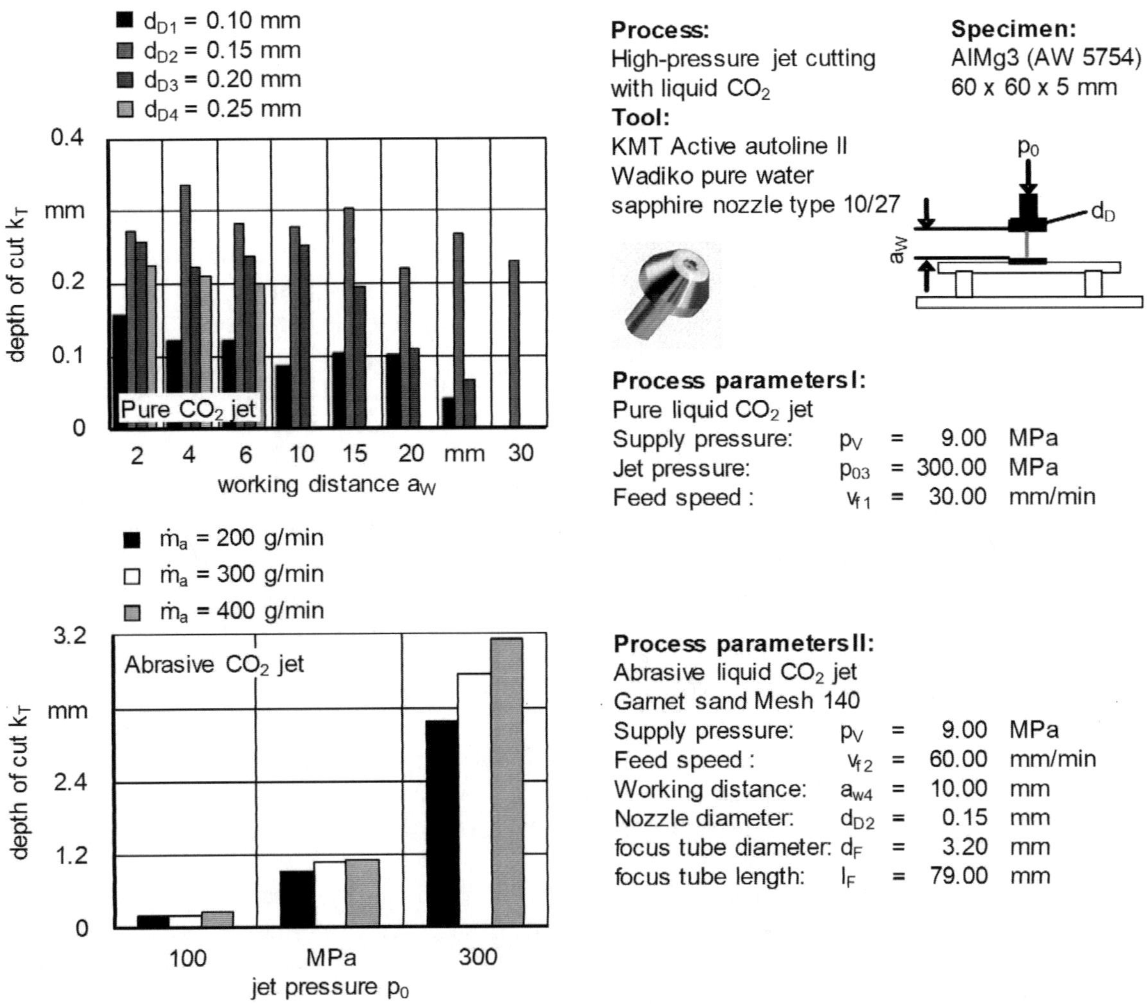

Figure 3. Depth of cut k_T for pure and abrasive CO_2 jetting of AlMg3.

The tests with garnet sand showed a significant increase in the depth of cut k_T compared to the pure jetting tests. The kerf also has a U-shaped cross-section with steep kerf flanks and a flat kerf base. The jet pressure p_0 and the nozzle diameter d_D have by far the largest influence on the maximum depth of cut k_T. Changing the pressure from $p_0 = 100$ to 300 MPa causes a 10-fold increase of the depth of cut k_T. Changing the nozzle diameter from $d_D = 0.10$ to 0.15 mm causes an increase in depth of cut by twice to three times. Contrary to expectations, the variation of the abrasive mass flow seems to have just small influence on the depth of cut k_T. However, it is determined that the increasing nozzle diameter and the mass flow results in an increased depth of cut. For example, the deepest kerfs were measured for all nozzle diameters at an abrasive mass flow of $\dot{m} = 400$ g/min and a pressure of $p_0 = 300$ MPa. With a nozzle diameter of $d_D = 0.15$ mm,

the deepest kerfs were measured at $k_T = 3.19$ mm. In the investigated area, the jet distance a_W has no strong influence on the depth of cut k_T. However, it strongly influences the kerf width k_B. For example, an increase of around 30 % was observed when the jet distance changed from $a_W = 5$ to 15 mm. Furthermore, it was determined that the kerf width is influenced together with the abrasive mass flow and jet pressure. A higher pressure straightens the jet and cuts about 10 % more narrow kerfs.
This assumption is also illustrated in Figure 1. It can be seen, that the jet is fraying at the edge and wave formation occurs. Only at increasing jet pressure, the inside of the jet can be identified as homogeneous liquid CO_2. By increasing the jet pressure from $p_0 = 100$ to 300 MPa the jet characteristics regarding cutting performance, homogeneity and jet expansion directly after the nozzle outlet could be influenced positively. As shown in Figure 1, the jet expands after the nozzle outlet and widens with an increasing distance. The pure jet diameter after the nozzle outlet is $d_{SI} = 1.08$ mm for the nozzle diameter $d_D = 0.1$ mm. This equates to an expansion by a factor of 10. The cutting performance at the workpiece only achieved a middle kerf width of $k_{BM} < 0.2$ mm, which corresponds approximately to the homogeneous liquid jet center. The expansion of the CO_2 is caused by the extreme pressure drop at atmospheric conditions and a concurrent temperature change directly after the nozzle outlet. Thus, the jet fluid CO_2 changes from liquid state to gaseous state at the outer edge of the jet.
The direct comparison between pure and abrasive jetting shows that by using the abrasive jetting process up to ten times deeper kerfs can be produced. The kerf width is also larger in abrasive jet cutting and exceeds the kerf width of the pure jet by up to 4.3 times. The ratio of k_B/k_{BM} is a reference for the kerf shape and consistently higher in pure jet cutting than in abrasive jet cutting. It amounts to pure jets up to 3.46 and in abrasive jets not more than 1.7. This explains the two different kerf shapes.
The proof that at low pressures and with the sharp-edged sand a low removal is possible, but nevertheless an acceleration of the particles is adjustable, qualifies the procedure for further blasting applications. An application of dry and highly accelerated shot peening is therefore conceivable, in which both small and large particles up to $d_G = 0.8$ mm with up to $v = 900$ m/s could solidify a metallic surface, comparable to previous work [3].

In additional investigations CO_2-microcrystals have been added to the jet and were processed with different jet feed speeds v_F and focus tube diameters d_F. As expected, increasing v_F and d_F results in decreasing depth of cut k_T (Figure 4). The feed speed has just marginal influence on the kerf width k_B. The main influence was proven for the focus tube diameter d_F, which naturally confines jet decay and centres the impact zone. The depth of cut is not competitive in comparison to the pure jet or the abrasive garnet, but the feasibility of a dry and residue-free abrasive jetting process was proven. Furthermore, the surface of the AlMg3 specimens showed unusual undirected craters (Figure 5c). Using the CO_2-microcrystals show just little depth of cut but give a hint to a potential for surface preparation, for example before coating processes. The reason are the different, jagged structures and undercuts in harmony with great roughness, as they are required for subsequent coating.
A comparison of the kerfs cut by the different jet fluids is shown in Figure 5, where the different length, quality and strength of the jets can be assumed. The investigation of the cutting surface shows a characteristic outcome for the pure waterjet: the depth k_T significantly increased and a rather U-shaped kerf profile was identified (Figure 5a). Quite different is the cutting result after jetting with liquid CO_2. The cutting surface shows a rather pore-like characteristic and the kerf

root is wavy and rough. These results indicate a significant difference at the behavior of the backwater between liquid CO_2 and water. The density difference and gaseous phase seem to manifest here. That indicates a multiphase state of the CO_2 jet, which means that the density and therefore the cutting performance are higher inside the jet (Figure 5b).

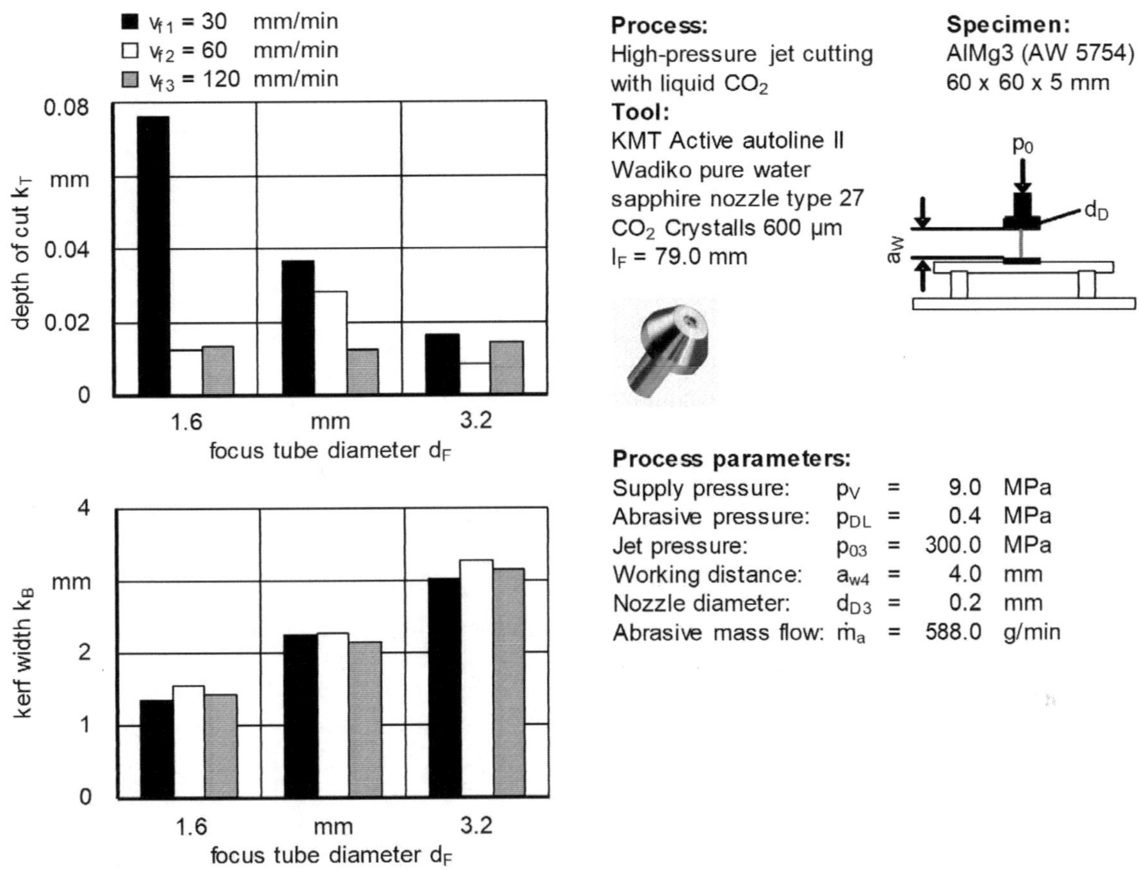

Figure 4. Depth of cut k_T and kerf width k_B for abrasive CO_2 jetting of AlMg3 with CO_2-microcrystals as abrasive.

The expansion of the CO_2 is caused by the extreme pressure drop at atmospheric conditions and a simultaneous temperature change directly after the nozzle outlet. Thus, the CO_2 changes from liquid state to gaseous state at the outer edge of the jet, which causes a density change of the CO_2. Alongside the jet from nozzle to workpiece, two different characteristics are influencing the decreasing cutting depth: On the one hand, there is a phase transformation from liquid to gas which causes a decrease in density of the fluid. On the other hand, the conical expansion of the jet enlarges the jet cross-sectional area and therefore decreases the effective area force. This distinct difference between water and liquid CO_2 jetting as well as pure and abrasive jetting enables a new technology for complex shapes and 3D-operations without jet catcher, for example necessary when machining using industrial robots.

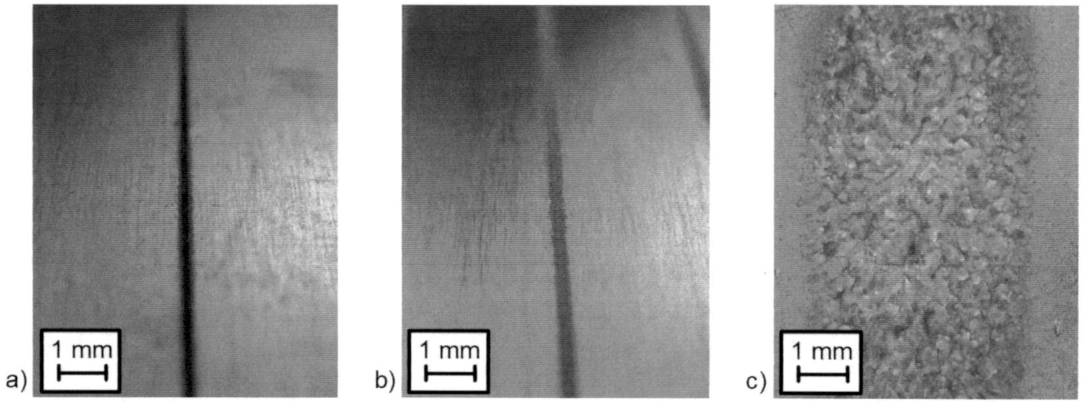

Figure 5. Cutting surface of AlMg3; a) pure water; b) pure liquid CO_2; c) liquid CO_2 with CO_2-microcrystals as abrasives.

3.2 The jet temperature

The temperature investigations were divided into pure jet tests and abrasive jet tests. In order to compare them with each other but not destroy the thermocouple, no abrasive was added, but the abrasive jet head was used. The jet pressure p_0, nozzle diameter d_D and jet distance a_w were varied. The analysis of the jet temperature T_S depending on the working distance a_W made it possible to trace the phase transition of the CO_2 jet. The temperature of the CO_2 jet depends significantly on the process parameters. To ensure that the CO_2 jet hits the workpiece in its liquid state, a temperature between $T_S = 31$ and -56.6 °C is required at jet pressures between $p_0 = 100$ and 300 MPa. As shown in Figure 6, the temperature is decreasing with higher distance a_W and lower jet pressure p_0 from $T_S = -14.7$ to -53.7 °C for the pure jet. Only at the abrasive jet and for the largest nozzle diameter, lowest pressure and highest distance to the workpiece, a temperature of $T_S = -58.3$ °C was measured. The abrasive jet expands to the gaseous phase, but is still able to accelerate abrasives as shown before.

The temperature T_S is beyond the brittle temperature T_V of certain plastics like Polypropylene (PP) and Polyvinylchloride (PVC). That could have an influence on cutting behaviour in certain applications and for some materials. For aluminium there should be no significant influence and at least no thermal damage. In most machining operations high temperatures insult the workpiece, the opposite behaviour is now proven for jetting with liquid CO_2.

By raising the nozzle diameter d_D the influence of a_W and p_0 is increasing. That could be an indicator, that the machine system has an influence. Higher volume flow rates require more pump strokes, which heats up the machine and the fluid. The heating of the system as a result of the high-pressure pump will overlap the effect of the phase transition. That could be another reason, why high jet pressures p_0 produce more stable CO_2 jets (Figure 1).

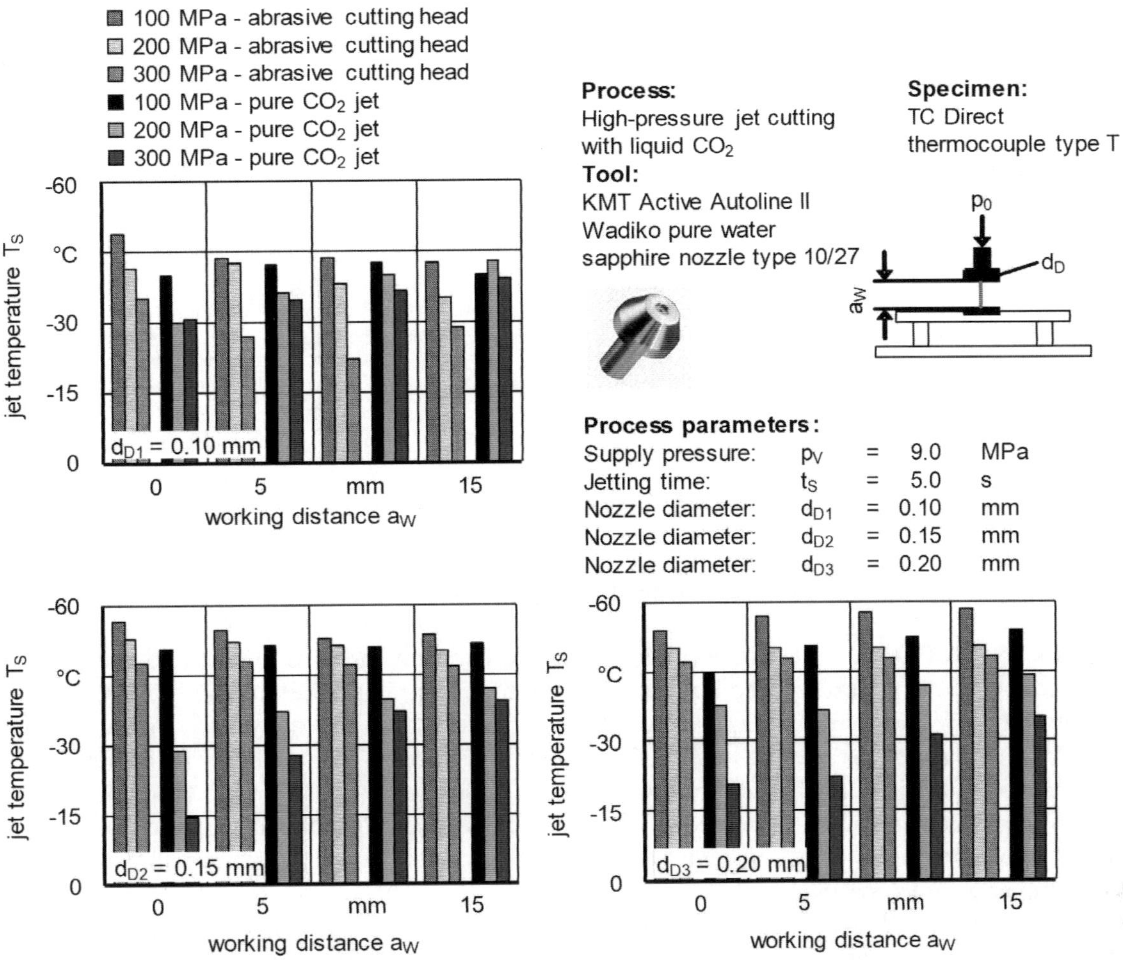

Figure 6. Jet temperature T_S of pure and abrasive liquid CO_2 jets.

In order to classify these findings, a temperature measurement at a conventional water jet machine system with a pure water jet was done. After a jetting time of $t_S = 30$ s, marked with a red line in Figure 7, jet temperatures of about $T_S = 78$ °C were observed. These temperatures could cause damage inside the workpieces, e.g. with technical plastics like Polypropylene (PP) with a melting point of $T_M = 80$ °C. The difference between liquid CO_2 and pure water jets is significant. Potential for CO_2 jet cutting arises in particular in applications where gentle, precise dry and cold machining is required, such as in precision engineering or electrical engineering. The processing of fibre composite materials, which are increasingly used in industrial applications, also offers great potential.

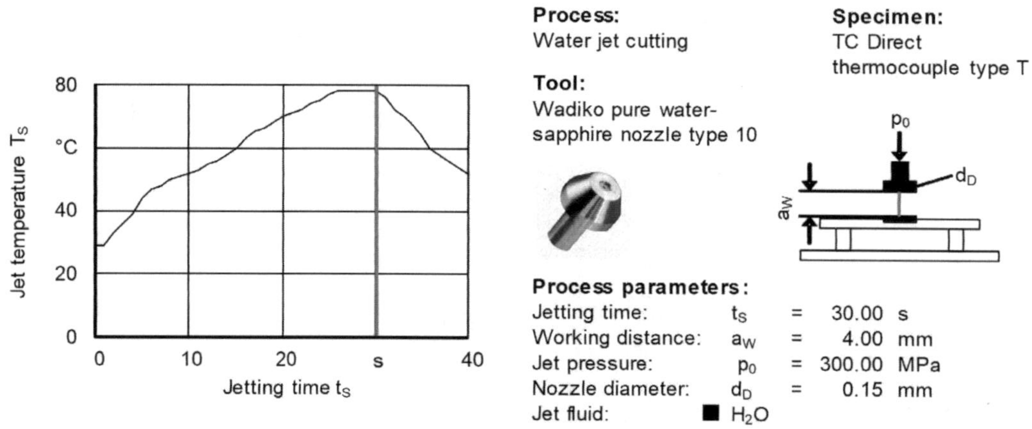

Figure 7: Jet temperature T_S of water jet cutting.

4 CONCLUSIONS

The described experimental and measurement setup provides a coherent and liquid high-pressure jet of CO_2 with and without abrasives which is comprehensible and reproducible. The experiments with specimens of AlMg3 have shown a general suitability of the process for a dry abrasive cutting process for various materials. The liquid CO_2 jet shows a similar behaviour to the water jet but with slightly lower kerf formation as a result of the different fluid density and machine system. The addition of abrasives into the liquid CO_2 jet and thus proof of the suitability for increasing the performance of the jet was shown. Solid CO_2 as an abrasive could not increase the cutting depth, but could produce unusual, undirected surface structures. Therefore, the acceleration of particles enables new potentials for opening up new applications, e.g. dry high-pressure shot peening or pretreatment prior to coating.

It could be shown, that the jet is fraying at the edge and a wave formation occurs, but with increasing jet pressure the jet is getting narrower and longer as well as the inside can be identified as homogeneous liquid CO_2. By increasing the jet pressure from $p_0 = 100$ to 300 MPa the jet characteristics regarding cutting performance, homogeneity and jet expansion directly after the nozzle outlet could be influenced positively.

The process can be optimized for a wide range of applications due to the possibility of influencing the jet properties by means of the process parameters. The phase conversion of the CO_2 jet can be used to machine workpieces within a certain temperature field.

However, another aspect that needs to be investigated is the effects on the workpiece surface depending on pressure and abrasive in order to widen the field of application. The targeted application of changing roughness and surface layer hardening would be a benefit.

5 ACKNOWLEDGMENTS

This paper is based on results acquired in the project DFG UH 100/163-3, which is kindly supported by the DEUTSCHE FORSCHUNGSGEMEINSCHAFT (DFG).

6 REFERENCES

[1] Momber, W. A.; Schulz, R.-R.: Handbuch der Oberflächenbearbeitung Beton – Bearbeitung-Eigenschaften-Prüfung. Basel, Birkhäuser, 2006.

[2] Kolb, M.: Wasserstrahlschneiden – Materialbearbeitung mit einem Hochdruckwasserstrahl. In: Bibliothek der Technik Band 295. München, Moderne Industrie, 2006.

[3] Uhlmann, E.; Faltin, F.: Feasibility Study of Waterjet Fine Particle Peening. MFN Science Update, March issue 2018, pp 56 - 58.

[4] Uhlmann, E. (Ed.); Bilz, M.: Marktstudie Kohlendioxidstrahlen 2010. Fraunhofer-Institut für Produktionsanlagen und Konstruktionstechnik, Berlin, 2010.

[5] Dunsky, C.; Hashish, M.: Cutting with high pressure CO_2 jets. Allen, A. G. (Hrsg.). Washington: QUEST Integrated, 1994.

[6] Bilz, M.: Möglichkeiten und Grenzen des Strahlspanens mittels CO_2-Hochdruckstrahlen. Berichte aus dem Produktionstechnischen Zentrum Berlin. Hrsg.: Uhlmann, E. Dissertation, Technische Universität Berlin. Stuttgart: Fraunhofer IRB, 2014.

[7] Engelmeier, L.: Flüssige, kohärente Kohlendioxidstrahlen zum Schneiden von Materialien in atmosphärischer Umgebung. Berichte aus der Verfahrenstechnik. Dissertation, Ruhr-Universität Bochum. Herzogenrath: Shaker, 2017.

[8] Uhlmann, E.; John, P.: High-pressure jet cutting with liquid CO_2. In: Proceedings of the 23rd international conference on Water Jetting in Seattle, USA. Ed.: BHR Group, 2016, pp 315 - 326.

[9] Uhlmann, E.; Sammler, F.; Richarz, S.; Heitmüller, F.; Bilz, M.: Machining of Carbon Fibre Reinforced Plastics. Procedia CIRP, Volume 24, 2014, S. 19 - 24.

[10] Uhlmann, E.; John, P.: Advancements of the manufacturing technology with high-pressure liquid CO_2 jets. In: Proceedings of the 2017 WJTA-IMCA Conference and Expo in New Orleans, USA. Ed.: WJTA®-IMCA®, 2017.

[11] Uhlmann, E.; John, P.: Dry cutting with high-pressure liquid CO_2 jets. Advanced materials letters, Volume 10, Issue 1, pp 02-08, 2019. DOI: 10.5185/amlett.2019.2231.

7 NOMENCLATURE

a_W	Working distance
a_{WS}	Workpiece edge lenght
d_D	Nozzle diameter
d_{WS}	Workpiece thickness
k_B	Kerf width
k_{BM}	Middle kerf width
k_L	Kerf length
k_T	Depth of cut
p_0	Jet pressure
p_V	Supply pressure
p_{DL}	Abrasive supply pressure
t_S	Jetting time
T_S	Jet temperature
v_f	Jet feed speed

2019 WJTA Conference and Expo
November 11-13, 2019 • New Orleans, Louisiana

Paper

Pulse Response and Structure Optimization of a Novel Disc-Type Pulse Jet Generating Tool

Shi Huaizhong, Hou Xinxu, Zhao Heqian
State Key Laboratory of Petroleum Resources and Prospecting, China University of Petroleum-Beijing, Beijing 102249, China

ABSTRACT

During drilling extended reach wells and horizontal wells, friction force can be reduced effectively by pulse tools like hydro-oscillator. Rotary percussion and torsional percussion can effectively improve the rock breaking efficiency. In order to provide a controllable pulse source, a disc-type pulsed water jet generator, driven by hydrodynamic impeller is proposed, whose flow area is changed by the rotation of discs. Therefore, the continuous jet is modulated into a pulse jet. A series of numerical simulations are performed to analyze the pulse characteristics of the generator. Specifically, the response characteristics of rotation speed at various flow rates, inner diameter of the impeller, blade helix angle, number of blades are calculated. In these cases, as the inner diameter of the impeller increases, the impeller rotating speed, which decreases first and then increases, varies from 500 rpm to 1200rpm. As the helix angle and flow rate increase, the impeller rotating speed increases significantly. When increasing flow rate of water from 20 L/s to 35 L/s, the pressure fluctuation amplitude increases to 0.7 MPa. The research results show that it can be realized to control the pulse characteristics by adjusting flow rate and structure parameters. This work originally provides the basis for parameters optimization and field application of this generator.

1. INTRODUCTION

Statistics show that the drilling time occupies nearly 1/3 of the well cycle (Anderson, E. E, et al. 1990), and increasing the rate of penetration（ROP）can significantly shorten the drilling cycle and reduce drilling costs eventually. Pulsed jet drilling technology can effectively utilize the bottom hole hydraulic energy, optimize the distribution of jet energy, improve the efficiency of rock-breaking and cuttings-cleaning, which will increase the rate of penetration. Another method, namely pulse vibration drilling technology, can introduce a vibration impact on the rock while rotating by using a pulse jet and a rotary shock, thus leading to a high ROP. For both the pulse jet drilling technology and the pulse vibration drilling technology, the pulse generator is required to provide a stable pulse power source to improve the rate of penetration ultimately.

To improve the ROP by pulsed power, Jack Kolle et al. (Kolle, Jack, and Mark Marvin.1999) developed a "Hydropusle" negative pressure pulse tool, which created a transient partial underbalance and a negative pressure pulse in the bottom hole by controlling the flow connectivity between the drill bit and the tool. Based on the transient flow principle (Johnson Jr, V. E., et al.1981), the LPMS-I pulse modulator used the mechanical oscillation blocking method to make the main valve of the tool periodically shut down the flow path of the drilling fluid, forming a negative pressure underbalance state near the bit, improving the efficiency of rock breaking. Another pulse cavitation jet generator designed by China University of Petroleum (Beijing) could generate a pulse by means of mechanical blocking, and enhance the pulse through the self-excited oscillation cavity. The pulse cavitation can be formed with the assistance of pulse cavitation nozzles(Li, Gensheng, et al. 2012; Fu, Jiasheng, et al. 2012; Shi, Huaizhong, et al. 2010; Li, Gensheng, and Shi, Huaizhong. 2008; Li, Gensheng, et al. 2009; Li, Gensheng, et al. 2005). In such a kind of pulse cavitation generator, the drilling fluid changed the flow direction when passing through a fluid guiding body. Then it impacted and pushed the impeller to rotate, changed the flow area, and generated speed fluctuation and pressure fluctuation of the drilling fluid, forming a pulse. Furthermore, the pulse could be amplified by the self-excited oscillatory cavity, and acted on the bottom of the well. Cui Longlian(Cui, Longlian, et al. 2012) et al. developed a frequency-adjustable pulse speed-up tool. When the drilling fluid impacted turbine blades, it lead the center drive shaft of the turbine to rotating and changing the flow area, which resulted in a periodic sealing by the periodical pulsed drilling fluid through the tool.

In this paper, a new kind of disc-type pulsed water jet generating device was proposed based on previous studies. A series of numerical simulation studies were carried out, including effects of impeller inner diameter, number of blades, blade helix angle and displacement on the dynamic characteristics of the impeller, as well as the effects of the speed of the dynamic disc valve, the diameter and number of the open hole, and the displacement on the pulse characteristics.

2. DISC-TYPE PULSED WATER JET GENERATOR

In this paper, a new kind of disc-type pulsed water jet generating device is described as shown in Figure.1, which consists of a power system and a pulse generating system. The power system includes an axial-rotating impeller, while the pulse generating system is composed of a dynamic disc valve and a static disc valve. The dynamic disc valve is connected to the impeller through the impeller shaft while the static disc valve is fixed together with the wall of the shell. Both the dynamic disc valve and the static disc valve are distributed with the same number of circular holes. When the drilling fluid passes through the impeller, it drives the impeller shaft to rotate, transforming the kinetic energy of the drilling fluid into the mechanical energy of the impeller shaft, and the rotating impeller shaft drives the dynamic disc valve to rotate. With the rotation of the dynamic disc valve, the circular holes on these two disc valves alternately overlap or disconnect, modulating the continuous drilling fluid flow into a pulse jet (Figure. 2). The advantage of this tool is that the amplitude and frequency of the hydraulic pulsed jet are easily controlled by adjusting structural parameters and drilling fluid parameters.

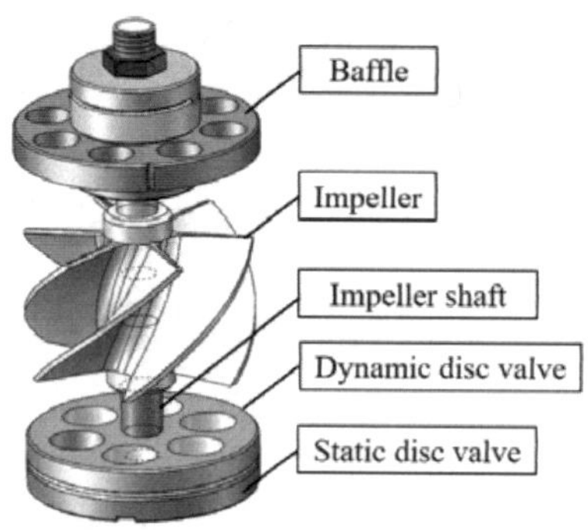

Figure.1 Novel Disc-type Pulsed Water Jet Generator

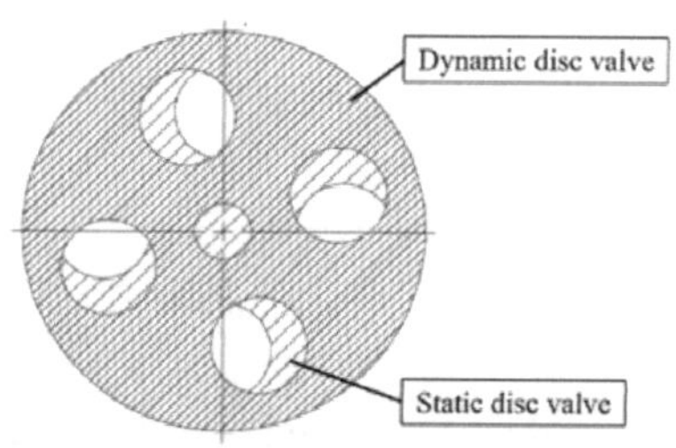

Figure.2 Mechanism Diagram of Pulsed Jet Flow

3. THEORETICAL ANALYSIS OF POWER SYSTEM

3.1 Basic Parameters

The impeller with right-pitch helical blades is used in this paper, whose outer diameter D_1 is 96mm and the blade thickness H is 4mm. The diameter D_0 of the sleeve through which the drilling fluid flows is 100mm, the radius R of the contact surface of impeller shaft and baffle is 5mm, and the coefficient of friction k is 0.1. In the working process, the blade generates torque, rotates under the impact of the fluid, transmits torque to the dynamic disc valve through the shaft, and drives the flow area to change. In this paper, pure water was used as the fluid medium, and the influence of structural parameters and drilling fluid parameters on the power output characteristics of the impeller motor was analyzed through theoretical calculation.

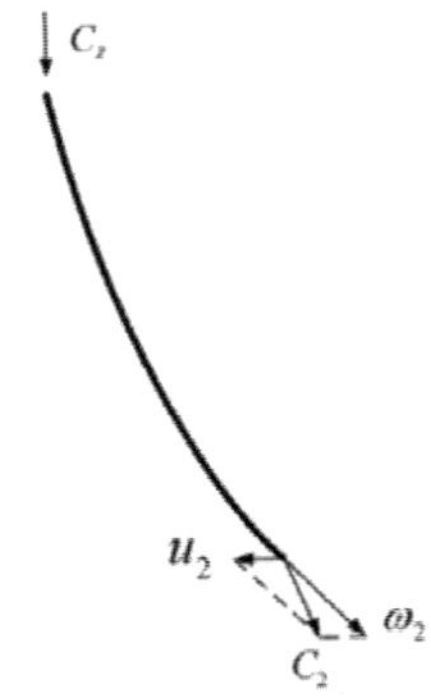

Figure 3. Diagram of Impeller Inlet and Outlet Velocity

3.2 Theoretical Calculation

According to vectors of impeller inlet and outlet velocity in Figure.3, The torque produced by liquid impinging on blades is given by:

$$M = \rho Q \frac{\bar{D}}{2}(C_z \sin\beta - u_2) \tag{1}$$

Where: ρ is the fluid density, kg • m^{-3}; Q is the flow rate through the impeller, L • s^{-1}; $\bar{D}$ is middle diameter of the impeller, m; C_z is actual axial velocity, m • s^{-1}; β is helix angle, rad; u_2 is circumferential velocity.

Outlet circumferential velocity is given by:

$$u_2 = \frac{\pi n \bar{D}}{60} \tag{2}$$

Considering that there is a 2mm gap between the blade and the sleeve wall, the actual flow through the blade should be relatively small, thus:

$$Q = aQ' \tag{3}$$

According to the relationship of the overcurrent area, a=0. 9293.

Axial velocity is given by:

$$C_z' = \frac{Q}{\pi \bar{D} b} \tag{4}$$

Since the blade has a certain thickness and number, the influence of the excretion coefficient on the axial velocity should also be considered, so the actual axial velocity is:

$$C_z = \frac{C_z'}{\varphi} \tag{5}$$

Excretion coefficient is given by:

$$\varphi = 1 - \frac{zc}{\pi \bar{D}} \tag{6}$$

Where: z is the thickness of blade, m; c is the number of blades, b is the radial length of the blade.

The weight of entire impeller and the drilling fluid impact force act on the action surface of the impeller shaft and the baffle, thus torque generated by friction is:

$$M_b = \int_0^R \frac{\mu\left[mg + \rho Q(C_z - C_2)\right]}{R^2} 2r^2 \mathrm{d}r = \frac{2\mu R\left[mg + \rho Q(C_z - C_2)\right]}{3} \tag{7}$$

Where: m is the weight of dynamic disc valve and blades, kg; C_2 is axial velocity of impeller outlet, m • s^{-1}, $C_2 = C_z \bullet \cos\beta$; μ is the friction coefficient between the impeller shaft and the baffle; R is the radius of contact surface between impeller shaft and baffle, m.

When the impeller shaft rotates at a constant speed, it can be considered to be in an equilibrium state, unite the formula (1) and formula (7):

$$n = \frac{60 C_z \sin\beta}{\pi \bar{D}} - \frac{120 M_b}{\pi \rho Q \bar{D}^2} \tag{8}$$

According to the above equation, the output parameters corresponding to different drilling fluid parameters and different structure parameters can be calculated, shown as Figure.4.

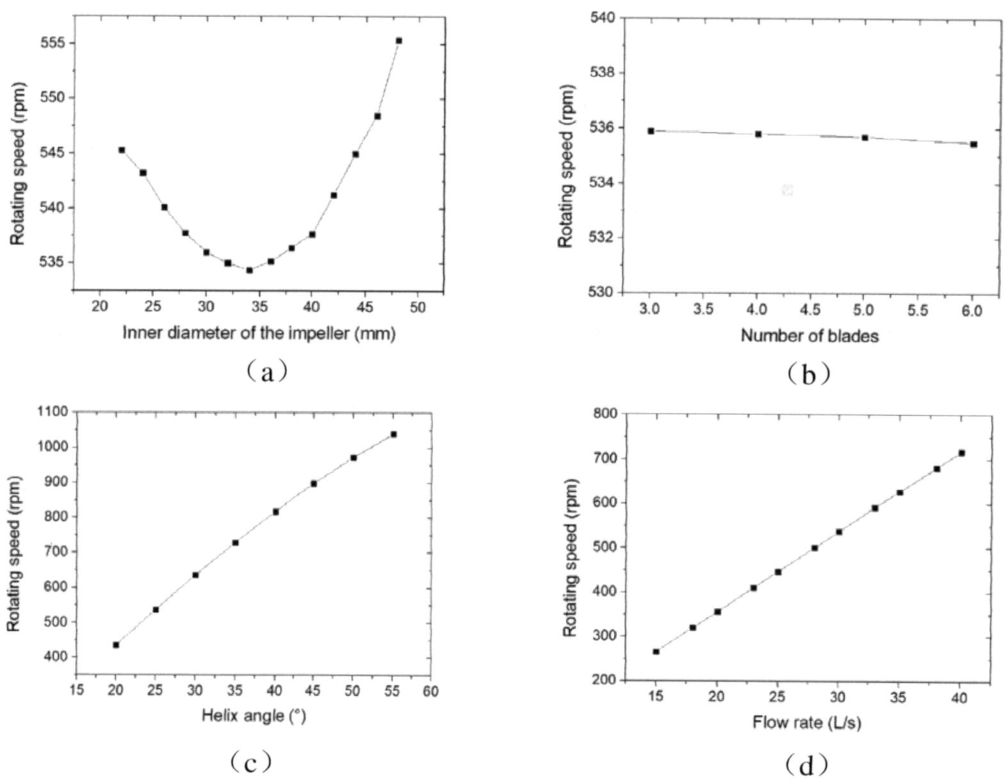

Figure.4 The Rotating Speed of the Impeller Varies with (a) Inner Diameter, (b) the Number of Blades, (c) the Helix Angle and (d) the Flow Rate of Drilling Fluid

According to the curves shown in Figure 4, as the inner diameter increases, the impeller rotation speed decreases first and then increases. On the other hand, the impeller rotation speed seems to be unaffected by the number of blades. Moreover, the rotation speed significantly increases with increasing helix angle and increasing flow rate.

4. NUMERICAL SIMULATION

4.1 Physical Model

For the impeller power system of the pulsed water jet generator, four kinds of inner diameter size, four kinds of blades, four kinds of helix angle impellers and six kinds of displacements were selected to complete the analysis of their effects on the torque, pressure loss and rotation speed, as shown in Table 1.

Table 1. Different numerical simulation parameters of impeller

No	Inner diameter(mm)	Number of blades	Helix angle(°)	Displacement ($L \cdot s^{-1}$)
1	28	3	25	15
2	34	4	35	20
3	40	5	45	25
4	46	6	55	30
5				35
6				40

For the pulse generation system, six kinds of rotation speeds, four kinds of displacements, five kinds of disc valve holes and four kinds of disc valve apertures are selected to complete the simulation analysis of the pulse characteristics, as shown in Table 2.

Table 2. Different numerical simulation parameters of disc valve

No	rotation speed (rpm)	displacement (L/s)	Number of holes	Diameter of holes (mm)	Drilling fluid
1	100	20	2	16	pure water
2	150	25	3	20	pure water
3	200	30	4	24	pure water
4	300	35	5	28	pure water
5	400		6		
6	500				

According to the above selected different impeller power system and pulse generating system, a three-dimensional model (shown in Figure.5) was built using SolidWorks, and imported into ICEM for meshing. Due to the complex structure of the impeller, the unstructured mesh was used for mesh division, and the structure of the pulse generating system was relatively regular, the mesh with better quality was used for mesh division. The number of meshes was 220,000 and 180,000 respectively, the mesh quality was greater than 0.6, and the orthogonal mesh quality was greater than 0.7, which was consistent with the mesh quality required for numerical simulation.

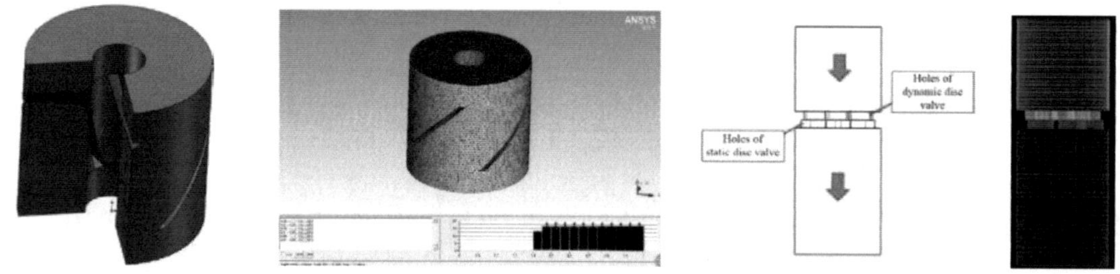

(a) (b) (c) (d)

Figure.5 Geometric Model and Structure Division of Impeller and Pulse Generating Mechanism

The Realizable k-ε model is selected, which was suitable for the description of rotating shear flow. The SIMPLE algorithm is a semi-implicit separation method and it belongs to the pressure correction method based on the finite volume method. It is the most widely-used flow field calculation method in engineering.

The inlet boundary is set as the velocity inlet while the outlet port is directly set as the atmosphere pressure. There is no slippery near the wall surface, and the wall function method is selected for the fluid phase. The clean water is used as the fluid phase, and the initial velocity of the fluid at the inlet boundary is given. Set the corresponding turbulent dissipation rate and turbulent kinetic energy. For the numerical simulation of the dynamic mechanism, the 6DOF (Six Degrees of Freedom) model in the dynamic mesh model is used to make the impeller passively rotate until the rotational speed is stable. For the numerical simulation of the pulse generating mechanism, the holes of the disc valves are given a certain rotational speed for active rotation.

4.2 Simulation Results and Discussion

4.2.1 Numerical Simulation Analysis of Impeller Power Mechanism

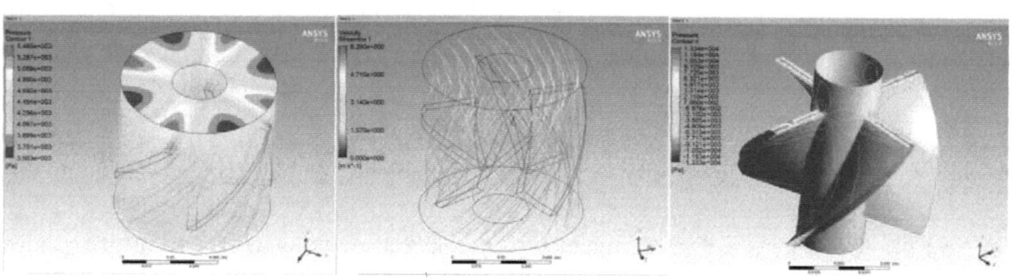

(a) (b) (c)

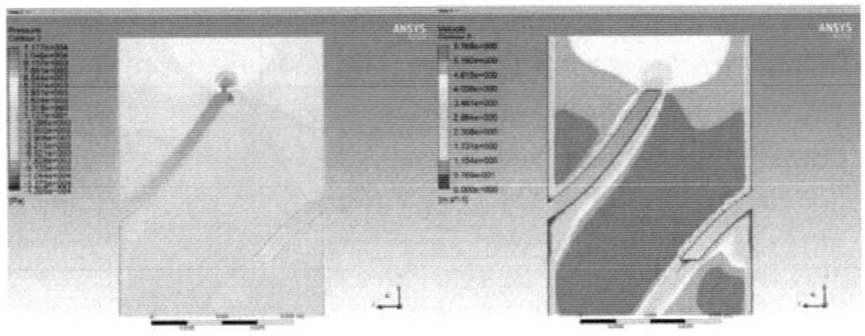

(d) (e)

Figure.6 Pressure and Velocity Distributions Near Impeller(a) Three-dimensional Pressure Distribution, (b) Flow Diagram, (c) Surface Pressure Distribution of Impeller, (d) Side View of Pressure Distribution, and (e) Side View of Velocity Distribution

It is found that the pressure is distributed like a petal at the entrance of the impeller through numerical simulation, as shown in Figure.6 (a), which should be closely related to the presence of the blade and increased pressure on the upper end surface of the blade. The flow rate of the drilling fluid also becomes larger as it enters the fluid channel between blades, and gradually remains substantially unchanged. At the same time, it can be seen from the figure that the flow rate of the drilling fluid becomes smaller or even zero near the wall surface of the blade, in accordance with the theoretical analysis of the influence of the wall on the flow velocity of the drilling fluid. The pressure exerted by the drilling fluid on the upper side of the blade is relatively large, so the resulting torque forces the impeller to rotate, converting the kinetic energy of the drilling fluid into the mechanical energy of the impeller.

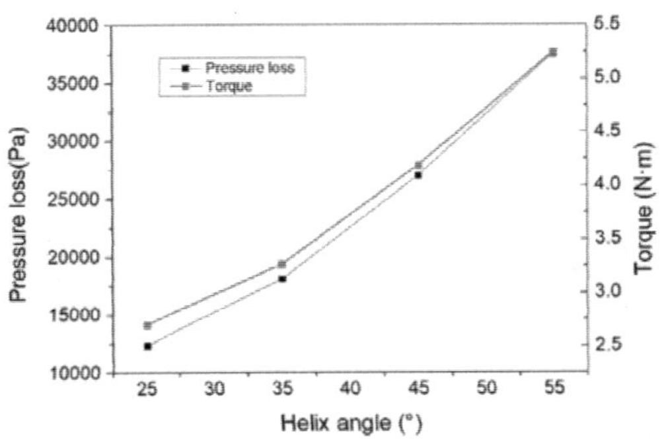

(a)

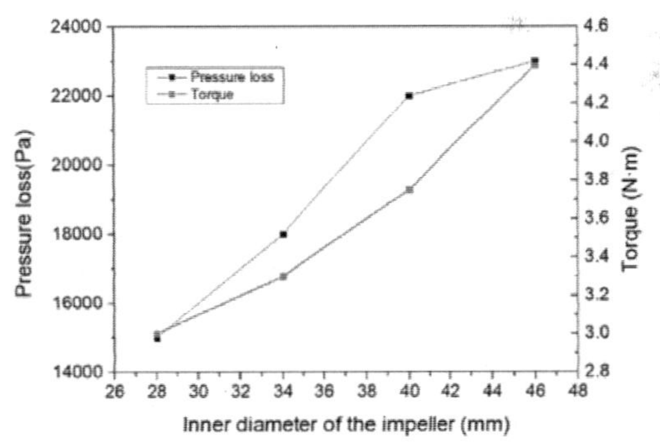

(b)

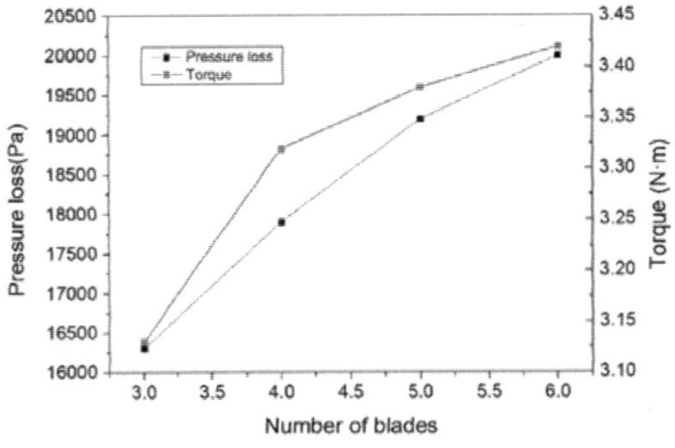

(c)

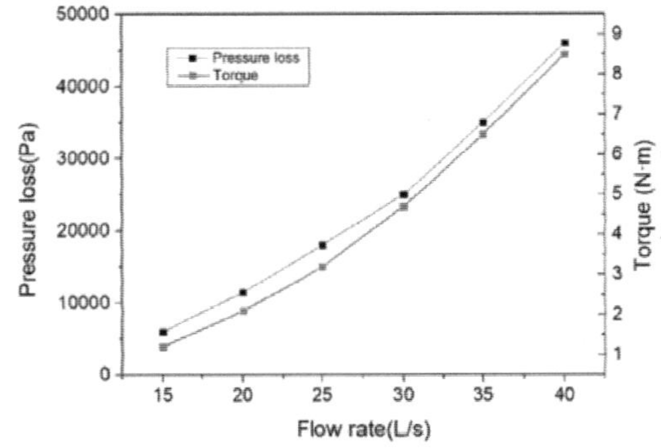

(d)

Figure.7 Changes of Pressure Loss and Torque with (a) Helix Angle of the Blade,(b)Inner Diameter of Impeller,(c) Number of Blades,(d) Flow Rate of Drilling Fluid

Figure 7 is a summary of the changes of pressure loss and torque as a function of blade helix angle, impeller inner diameter, number of blades, and drilling fluid displacement. It can be seen from Figure. 7(a) that both the pressure loss and the torque increase with the increase of the helix angle of the impeller. When the helix angle is increased from 25° to 55°, the torque is increased by 94% and the pressure loss is increased by 204%. The final circumferential velocity of the drilling fluid becomes larger as the helix angle become larger. According to the momentum conservation, the torque acting on the impeller is greater.

According to Figure. 7(b), both the pressure loss and the torque increase as the inner diameter of the impeller increases. When the inner diameter was increased from 28 mm to 46 mm, the torque and pressure loss is increased by 47% and 53%, respectively. The average radius of the drilling fluid acting on the impeller increases as the inner diameter of the impeller increases, so the same pressure produces a greater torque. At the same time, as the flow area becomes smaller, the flow rate increases and the pressure loss increases.

It can be seen from Figure. 7(c) that both the pressure loss and the torque increase with the increase of the number of blades, but the variation range is relatively small, which are 23% and 9%, respectively. As the number of blades increases, the flow area and pressure loss increase due to the blade having a certain thickness.

It can be seen from Figure. 7(d) that the pressure loss and torque increase with the increase of the drilling fluid displacement, and the increment speed becomes more and more obvious. With the displacement increases from 15L/s to 40L/s, the pressure loss and torque get increased by 660% and 608%, respectively.

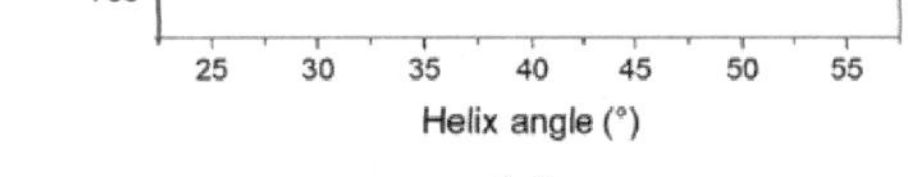

(a)

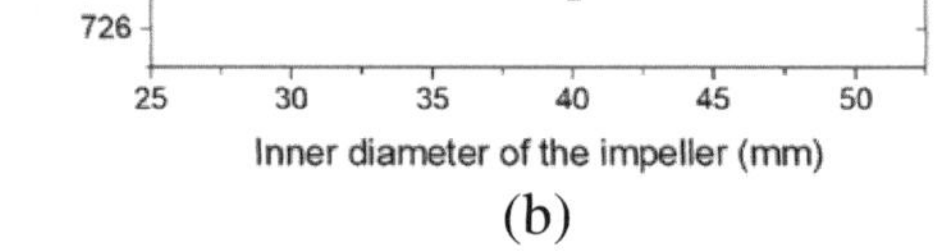

(b)

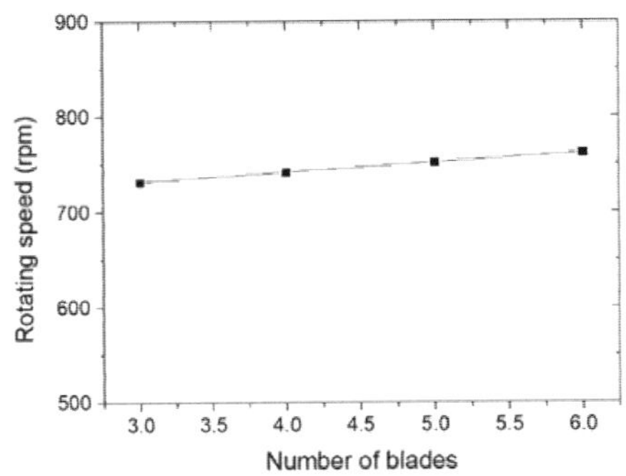

(c)

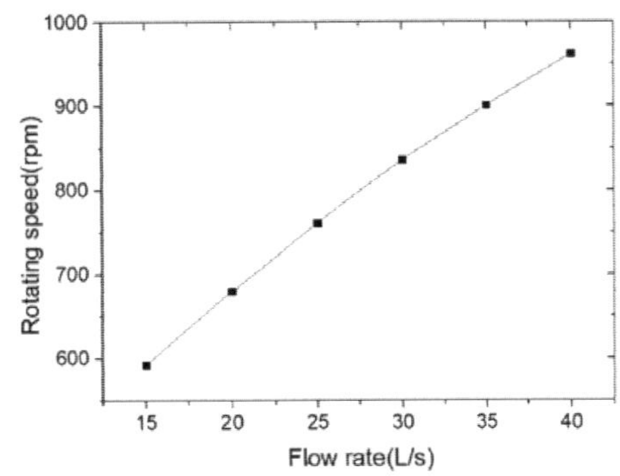

(d)

Figure.8 Rotating Speed Changes with (a) Helix Angle of the Blade,(b) Inner Diameter of Impeller,(c) Number of Blades,(d) Flow Rate of Drilling Fluid

Figure 8 is a graph showing the changes of the rotating speed with the impeller helix angle, the inner diameter of the impeller, the number of blades and the displacement of the drilling fluid. It can be seen from those curves that the rotary speed of the impeller significantly increases with increasing helix angle and increasing flow rate. However, increasing the number of blades seems to have few effects on improving the rotary speed. Moreover, there exists a minimum rotary speed along with increasing inner diameter of the impeller, which may be related to the interaction of the inertia and the torque.

These numerical simulation results corresponds well with the theoretical results shown in Figure 4. As for the small difference in the data, the numerical simulation process does not take into account the blocking effect of the dynamic disc valve on the drilling fluid, so that more energy from the drilling fluid is converted into mechanical energy of the impeller.

4.2.2 Numerical Simulation Analysis of Pulse Generating Mechanism

In a pulse generation cycle, the pressure mapping and velocity mapping at different times are shown in Figure. 9 and 10, respectively. It can be seen from the figure that when the position of the holes of the dynamic disc valve is shifted away from the holes of the static disc valve, the pressure in the holes of the dynamic disc valve rapidly increases, and the speed drops sharply. Again, when the position of the holes of the dynamic disc valve coincides with the position of the holes of the static disc valve, the pressure is reduced and the speed tends to be high. The continuous rotation of the dynamic disc valve causes the overcurrent area to change periodically, resulting in fluctuations in pressure and speed. According to the numerical simulation results, the influence of each parameter on the amplitude, frequency and pressure loss of pressure pulse is analyzed.

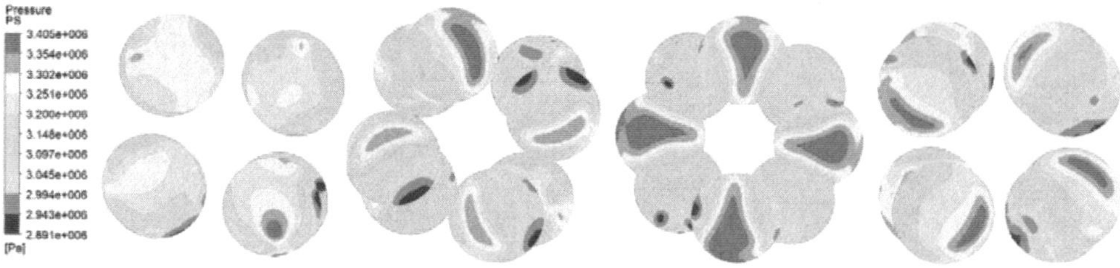

Figure.9 Pressure Distribution of the Pulse Generation Mechanism

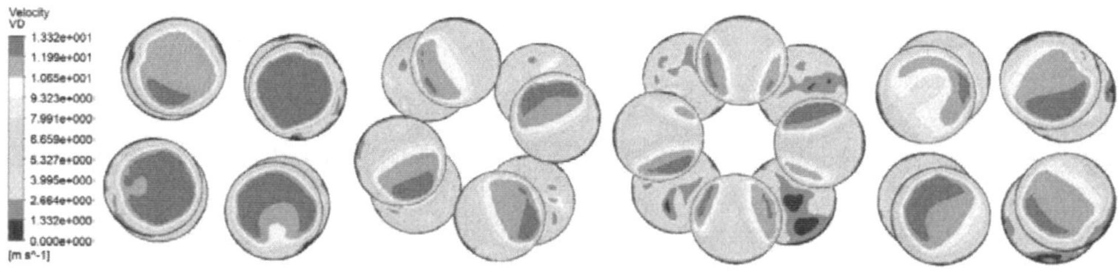

Figure.10 Velocity Distribution of the Pulse Generation Mechanism

Figure 11 shows the relationship between the pressure loss and the parameters of the pulse generating mechanism. As shown in Figure. 11 (a), the rotation speed does not cause a change in the pressure loss since the rotation speed does not affect the flow area. As shown in Figure. 11(b), when the displacement increases, the flow rate of the drilling fluid increases, and the pressure loss gradually increases. As shown in Figure. 11 (c) and Figure. 11 (d), the pressure loss is gradually reduced due to the increased flow area, which is caused by an increase in the number of holes and the diameter of the holes.

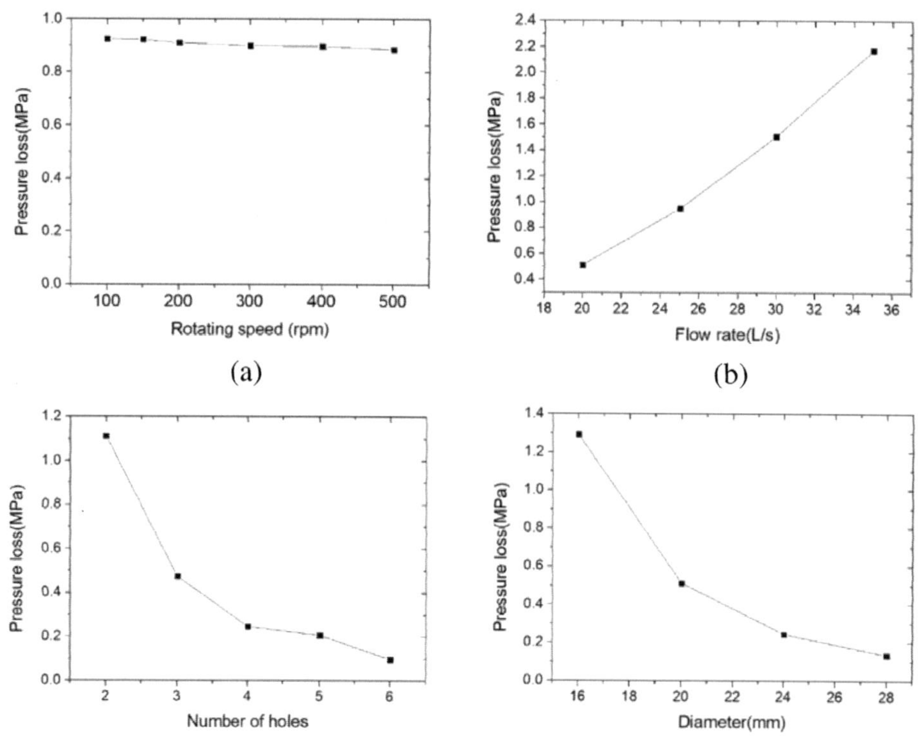

(c) (d)

Figure.11 Pressure Loss Changes with (a) Rotating Speed, (b) Flow Rate, (c) Number of Holes, (d) Diameter of Holes

The pulsation amplitudes of dynamic pressure, static pressure and total pressure are counted and plotted in Figure.12. When the rotation speed increases, the pulsation amplitude gradually decreases, and the pulse effect also decreases. The increase of the rotation speed causes the overcurrent area to alternate. When the frequency of the pulsation is higher than the propagation speed of the pulse effect, the pulse effect is weakened, resulting in a slow response of the fluid to the pressure and a decrease in the pulsation amplitude.

Figure. 12(b) shows that as the displacement increases, the flow rate increases, the pulse effect is gradually enhanced, so the pulsation amplitude increases. According to the curve, the relationship between the pulsation amplitude and the displacement can be well fitted by quadratic equations, whose coefficients of correlation are close to 1.

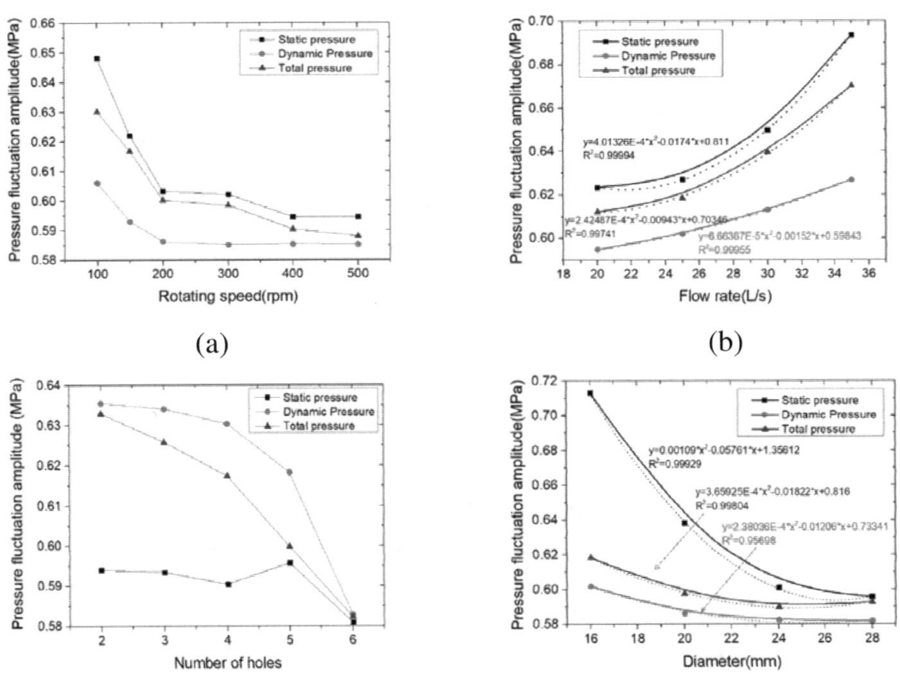

(a) (b)

(c) (d)

Figure.12 Pressure Fluctuation Amplitude Changes with (a) Rotating Speed, (b) Flow Rate, (c) Number of Holes, (d) Diameter of Holes

Figure. 12(c) and Figure. 12(d) respectively describe the influence of the number and the diameter of the holes on the amplitude of the pulsation. When the diameter of the holes is constant, increasing the number of holes brings increased flow area and thus result in a decrease in the pulsation amplitude. On the other hand, when the number of holes of the disc valve is constant, increased diameter of the holes of the disc valve also causes an increase in the flow area, thereby causing a decrease in the amplitude of the pulsation. The effect of the diameter of the holes on the amplitude of the pulsation can also be well fitted using a quadratic function.

As shown in Figure 13, the effects of each parameter on the pulse frequency are also drawn out. Along with increasing rotation speed, the pulse frequency linearly increases. However, the displacement and the diameter have no significant effects on the pulse frequency, which is showed Figure 13 (b) and Figure 13 (d). The pulse fluctuation frequency generated by the pulse generating device is mainly affected by the variation law of the flow area. Since the displacement and the diameter of the holes do not affect the change frequency of the flow area, there is no influence on the pulse frequency of the pulse generating device. The higher the rotation speed, the higher the frequency of change of the overcurrent area, resulting in an increase in the frequency of the fluctuation. As shown in Figure.13(c), when the number of holes of the disk valve increases, the frequency of the pulse fluctuation increases, because the relative distance of the disk valve hole becomes smaller as the number of holes increases. As the dynamic disk valve rotates, the overcurrent area changes more and more quickly, thereby causing an increase in the pulse frequency.

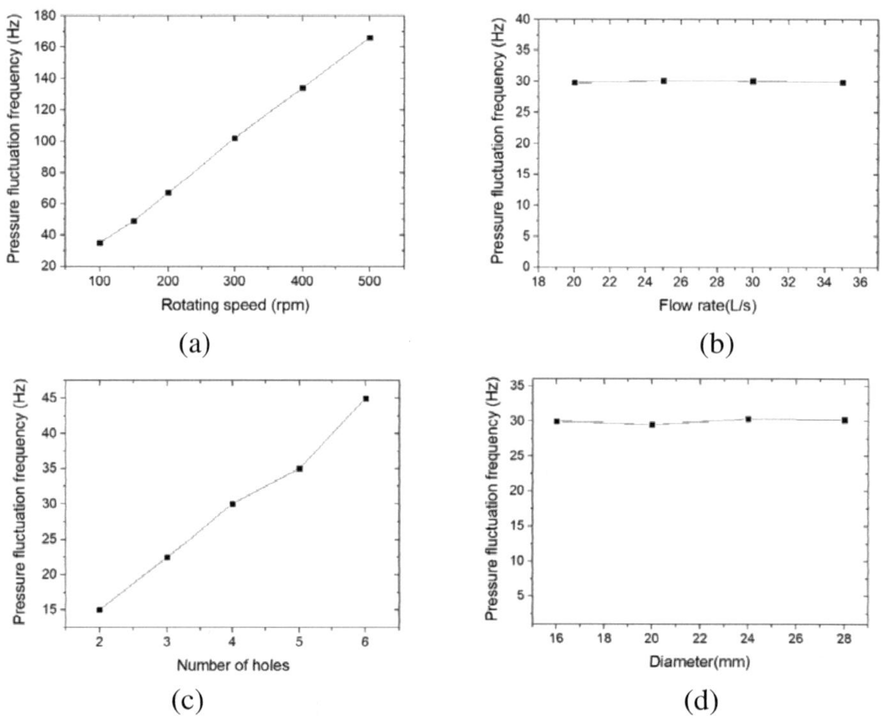

Figure.13 Pressure Fluctuation Frenquency Changes with (a) Rotating Speed, (b)Flow Rate, (c) Number of Holes, (d) Diameter of Holes

5. CONCLUSION

In this paper, a disc-type pulsed jet generating device was proposed, which used the impeller to drive the dynamic disc valve to rotate, and modulates the continuous drilling fluid flow into a pulse jet to act on the bottom hole. The theory combined with the numerical simulation analysis method is used to analyze the influence of the structural parameters of the power mechanism and the pulse generating mechanism. By comparing the cases set in this article, we found that:

(1) The rotation speed of the impeller varies from 500 to 1200 rpm. In general, the rotation speed significantly increases with increased helix angle and increased flow rate. However, it seems to be little affected by the number of blades and it should has a minimum value when the inner diameter of the impeller falls in a range of 20 to 50 mm.

(2) Both the pressure loss and the torque increase with the impeller helix angle, the inner diameter of the impeller, the number of blades and the displacement of the drilling fluid. Specifically, when the helix angle is increased from 25° to 55°, the torque is increased by 94%, and the pressure loss is increased by 204%. When the displacement is increased from 15L/s to 40L/s, and the pressure loss and torque are increased by 660% and 608%, respectively.

(3) For the pulse generating mechanism, the change of the rotation speed does not cause the change of the pressure loss; As the displacement increases, the pressure loss increases linearly; And as the number of holes and the diameter of the holes increase, the flow area increases and the pressure loss gradually decreases.

(4) The pulsation amplitude increases with increased flow rate and decreased values of rotation speed, the hole diameter and the number of holes. Especially, the effects of the flow rate and the hole diameter on the pulsation amplitude can be well descripted by quadratic functions.

(5) The pulse frequency significantly increases with increased rotation speed and increased number of holes. Nevertheless, there is no significant effects of the displacement and the hole diameter on the pulse frequency.

(6) To achieve a high pulse amplitude and a low pulse frequency, the disc-type pulse jet generating tool should have a small impeller helix angle, an inner diameter of the impeller within (30mm, 40mm), a small number of disc holes, and a small hole diameter. Considering the minimization on the press loss through the tool, the number of disc holes is suggested to be 4 while the hole diameter should be about 24 mm.

ACKNOWLEDGMENTS

Sincere thanks to National Major Scientific Research Instrument Development Project of China (No. 5182780013). This support is gratefully acknowledged by the authors, who are also grateful to the reviewers of this paper for their detailed comments.

REFERENCES

Anderson, E. E, et al. "Deep Drilling Basic Research: Volume 4 - System Description. Final Report, November 1988--August 1990." (1990).

Kolle, Jack, and Mark Marvin. "Hydropulses increase drilling penetration rates." Oil & gas journal 97.13 (1999): 33-37.

Johnson Jr, V. E., et al. Feasibility study of tuned-resonator, pulsating cavitating water jet for deep-hole drilling. No. SAND-81-7126. Sandia National Labs., Albuquerque, NM (USA); Hydronautics, Inc.,

Laurel, MD (USA), 1981.
Li, Gensheng, et al. "Hydraulic Pulsed Cavitating Jet-Assisted Drilling." Petroleum Science and Technology 27.2(2009):197-207.
Fu, Jiasheng, et al. "A novel tool to improve the rate of penetration--hydraulic-pulsed cavitating-jet generator." SPE Drilling & Completion 27.03 (2012): 355-362.
Shi, Huaizhong, et al. "Improving the rate of penetration by hydraulic pulsating-cavitating water jet under-balance pressure drilling." Petroleum Exploration & Development 37.1(2010):111-115.
Li, Gensheng, and Shi, Huaizhong. "Mechanisms and tests for hydraulic pulsed cavitating jet assisted drilling." Petroleum Exploration and Development 35.2 (2008): 239.
Li, Gensheng, et al. "An experimental study on hydraulic pulsed cavitation jet drilling in deep wells." 2009 American WJTA Conference and Expo. 2009.
Li ,Gensheng, et al. "Advances in Investigation and Application of Self-resonating Cavitating Water Jet." Engineering Science (2005).
Cui, Longlian, et al. "Development and Application of Adjustable Frequency Pulse Jet Generating Tool to Improve Rate of Penetration in Deep Wells." IADC/SPE Asia Pacific Drilling Technology Conference and Exhibition. Society of Petroleum Engineers, 2012.

NOMENCLATURE

D_1 Outer diameter of the impeller, mm;
D_0 Diameter of the sleeve, mm;
H Thickness of the blade, mm;
R Radius of the contact face, mm;
K Coefficient of friction;
ρ Fluid density, $kg \cdot m^{-3}$;
Q Flow rate, $L \cdot s^{-1}$;
$\overline{D}$ Middle diameter of the impeller, m;
C_2 Axial velocity, $m \cdot s^{-1}$;
C_z Actual axial velocity, $m \cdot s^{-1}$;
u_2 Circumferential velocity , $m \cdot s^{-1}$;
β Helix angle, rad;
ψ Excretion coefficient;
z Thickness of blade, m;
c Number of blades;
b Radial length of blade, m;
m Mass, kg;
μ Friction coefficient;
M Torque, $N \cdot m$;
M_b Friction torque, $N \cdot m$;
N Rotating speed, rpm.

2019 WJTA Conference and Expo
November 11-13, 2019 • New Orleans, Louisiana
Paper

Detection method and application of a self-resonating water jet based on the flow signal within the pipeline

Tengfei Cai, Yan Pan, Fei Ma

University of Science and Technology Beijing

Beijing 100083, China

ABSTRACT

The self-resonating water jet is characterized by the high-frequency pressure oscillation and strong cavitation. Accurately grasping the jet characteristics is a prerequisite for the application research of self-resonating water jet. In this paper, a detection method for self-resonating water jet characteristics based on the pressure signal within the pipeline was studied. The pressure sensors were arranged in the front pipeline outside the high pressure tank to avoid the influence of high ambient pressure. The test results show that the spectral characteristics acquired from the pressure signal in the pipeline agree with the signal in the chamber and theoretical calculations results. Furthermore, this detection method was successfully applied for studying the influence of nozzle lip geometry on the characteristic of self-resonating water jet.

Organized and Sponsored by WJTA®

1 INTRODUCTION

Self-resonating water jet is a kind of high-efficiency jet with the advantages of cavitation jet and pulse jet, which can significantly improve the erosion effect under high confining pressure environment. So the self-resonating water jet technology has a promising prospect in deep-sea mining, submerged cleaning and petroleum drilling(Johnson et al. 1984 and Li et al. 2005).

The characteristics of the self-resonating water jet – high frequency pressure oscillation and strong cavitation - directly determine its erosion effect. Therefore, accurately grasping the characteristics of the self-resonating water jet is the prerequisite for the application and research of self-resonating water jet. The traditional detection methods for the characteristic of self-resonating water jet mainly include the strike test method and detection of the pressure signal in the nozzle cavity. The strike test method is to add a sensor on the target plate to detect the oscillating characteristics of the self-resonating water jet(Chahine et al. 1983, Hu et al. 2015, Li et al. 2016). However, this method indirectly acquires the pressure oscillation characteristics of the self-resonating water jet through the target vibration, it has strong interference and low precision. The signal detection method in the nozzle cavity acquires the fluid signal directly from the nozzle(Chahine et al. 1983, Ma et al. 2016). The signal is very powerful with less disturbance and high precision, but the sensor arrangement is difficult especially for the small-sized nozzle. Besides, both detection methods are unable to overcome the influence of high confining pressure environment. As the confining pressure increases, the reliability of the sensor and target is significantly reduced and the effectiveness is drops dramatically.

To solve the above problems, this paper proposes a method for detecting the characteristics of self-resonating water jet based on pressure signal propagating within pipelines. The pressure sensor was installed on the pipeline right before the nozzle outside the high pressure tank, to acquire the pressure signal resulting from the self-resonance.

2 EXPERIMENT

2.1 Facilities

The self-resonating water jet is usually applied in a confining pressure environment, and the advantages of the technology can only be fully exerted under the confining pressure condition. For this purpose, a test device with high confining pressure conditions is built, as shown in Figure 1. The high pressure tank is used to simulate a high confining pressure environment. The confining pressure is controlled by the relief valve, and the test nozzle is placed in the high pressure tank. The adjustment of the incoming pressure, the confining pressure and the target distance can be realized by the control cabinet. The pressure sensor installed on the wall is used to collect the confining pressure parameter, and the flow rate is recorded by the rotameter.

There are 3 measuring points in the test device. The measuring point 1 and the measuring point 2 are used for collecting the pressure signals inside the pipeline, which are respectively arranged near the outlet of the high pressure pump and on the pipeline right before the high pressure tank, as shown in Figure1. Measuring point 3 is used to collect the signals from the pressure sensor arranged on the side wall of the nozzle inside the high-pressure tank as a comparative test. The output of transducers was captured with the data logger (Model: LMS SCADAS Mobile SCM05) which is connected to the computer. The time-domain and frequency-domain characteristics can be obtained in real time.

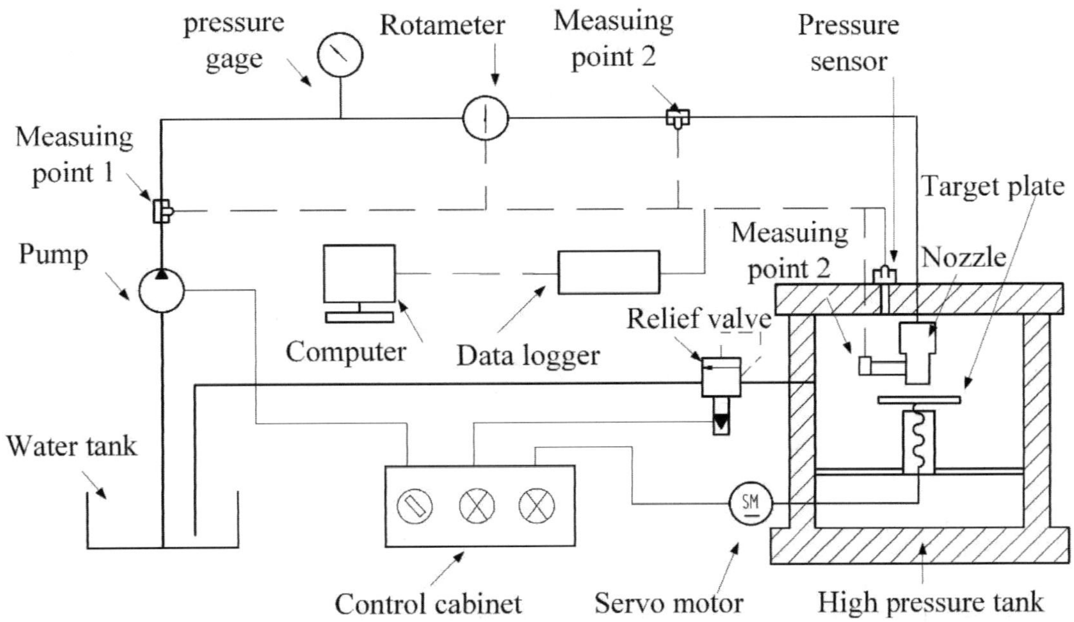

Figure 1. Schematic of the experimental setup

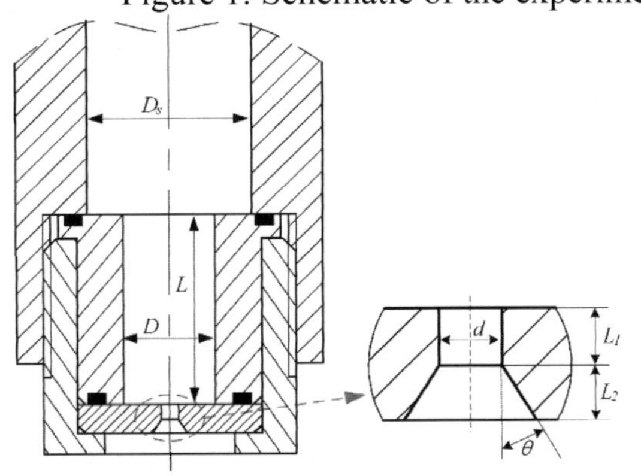

Figure 2. Schematic diagram of the organ-pipe nozzle and lip geometry.(a) Organ-pipe nozzle; (b) Enlarge of the nozzle lip

2.2 Nozzle geometry

Due to its strong erosion ability, low flow resistance coefficient and simple structure, the organ -pipe nozzle has been extensively studied and applied, thus the organ-pipe nozzle was used to produce self-resonating water jet in this experiment. A schematic of the organ-pipe nozzle is shown in Figure 2. Organ-pipe nozzle consists of an upstream area contraction (Ds/D), a downstream area contraction (D/d), and a resonant chamber with a length of L and a diameter of D. Peak resonance will occur when the acoustic natural frequency of the organ-pipe is close to the self-excited frequency(Johnson et al. 1984).

The self-excited frequency f^*, corresponds to the Strouhal number S_d , can be defined as

$$f^* = S_{\mathrm{d}} \frac{v}{d} \tag{1}$$

where, d and v are the diameter and velocity of the jet individually.

Organ-pipe's acoustic natural frequency f^n is determined by the chamber length L which is expressed below:

$$f^n = K_n \frac{c}{L} \tag{2}$$

For $Ds/D > 1$, the "mode parameter" K_n is given by

$$K_n = \begin{cases} n/2\,, & \frac{D}{d} < 1/\sqrt{\mathrm{M}} \\ (2n-1)/4\,, & \frac{D}{d} > 1/\sqrt{\mathrm{M}} \end{cases} \tag{3}$$

In these expressions n is the mode number of the organ-pipe, M stands for the Mach number(Johnson et al. 1984).

The major challenge in designing an organ-pipe nozzle is that the initial self-excited frequency has to match with the chamber's acoustic natural frequency. This can be described by the following equation:

$$f^* \approx f_n \Rightarrow L/d = \frac{K_n}{MS_d(1+\beta)} \tag{4}$$

Based on the above principles, the organ-pipe nozzle was designed, and the parameters are listed in Table 1.

Table1 Test parameters and self-resonating frequency

D_s(mm)	D(mm)	d(mm)	L(mm)	f^n(kHz)
23	10	2	24	14.9

As shown in Figure 2(a), the self-excited nozzle is constituted by the organ-pipe chamber and the orifice. A disk formed nozzle tip is bored the cylindrical orifice of diameter d, length L_1, and a conical expansion section of length L_2, angle θ (Figure 2(b)). In this study, the influence of nozzle lip geometry on the self-resonating water jet characteristic is studied by the above detection method. Table 2 lists the geometric properties of the nozzle lips investigated. The velocity of the jet is kept constant during the test, and the test parameters are shown in Table 3. At the flow rate, the motor speed is 621 r·min^{-1}, and the gearbox reduction ratio is 4.45. The pressure pulsation fundamental frequency f_p caused by the three-piston high-pressure pump is 6.98 Hz.

Table2 Geometry of the nozzle lips that employed

L_1/d	L_2/d	θ(°)
15	1.0, 1.5, 2.0, 2.5, 3.0	20

Table3 Test parameters and self-resonating frequency

P(MPa)	v(m.s^{-1})	c(m.s^{-1})	f_p(Hz)
15	165	1480	6.98

3 ANALYSES OF EXPERIMENT RESULTS

3.1 Pressure oscillation characteristics

To validate the detection method, pressure oscillation characteristics as the main characteristic of self-resonating jet, which are usually described by the time-domain and frequency-domain characteristics, were uncovered firstly.

3.1.1 The time-domain characteristics

The measuring points 1 and 2 are selected for analysis. The test parameters are shown in Table 2. During the test process, the confining pressure and target distance parameters are adjusted to achieve peak resonance. The confining pressure is 1.6 MPa and the target distance is 3*d*. The time-domain pressure signals of the measuring points 1 and 2 are obtained as shown in Figure 3.

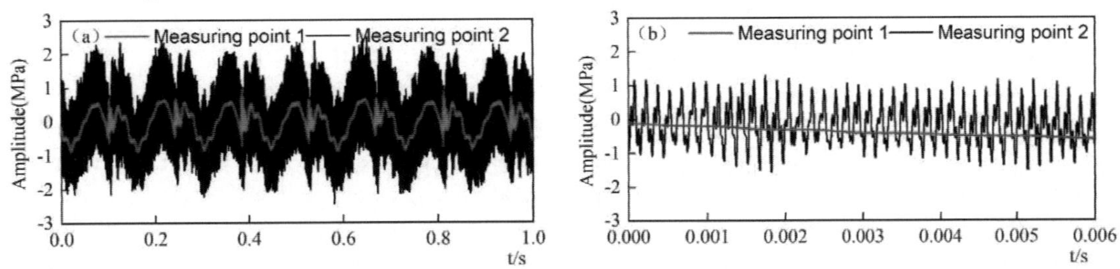

Figure 3. (a)Time-domain waveform of the pressure signal and (b) the locally amplified time-domain waveform

It can be seen from Figure 3(a) that the signal of the measuring point 1 is a low-frequency pressure fluctuation with an amplitude of about 0.8 MPa, but the signal amplitude of the measuring point 2 is about 2.0 MPa and contains the frequency component of point 1. It is also clear from the Figure 3(b)(the locally amplified time-domain waveform) that the signal of the measuring point 1 is approximately a straight line within 6 ms, showing a low-frequency characteristic, while the signal of the measuring point 2 exhibits obvious high-frequency periodic characteristics.

The measuring point 1 is close to the pump outlet, and the collected signal mainly comes from the low-frequency pressure pulsation caused by the reciprocating motion of the plungers and the random interference of the pipeline. The measuring point 2 is close to the test nozzle, and the signal acquired is mainly caused by high-frequency pressure oscillations and intense cavitation noise. Based on the acoustic theory, the high-frequency oscillation signal of measuring point 2 propagates upstream along the pipeline fluid, and the low-frequency pulsation signal of measuring point 1 propagates downstream. While the attenuation coefficient decreases as frequency increases. So the signal of measuring point 2 contains the component of the measuring point 1, while the signal of the measuring point 1 has almost no component of the signal 2. The time-domain pressure signal analysis results show that the pressure signal in the pipeline can clearly show the high-frequency pressure oscillation characteristics of the self-resonating jet. Nevertheless, the composition of the oscillating signal cannot be identified, and the pressure oscillation signal must be analyzed in the frequency domain.

3.1.2 The frequency-domain characteristics

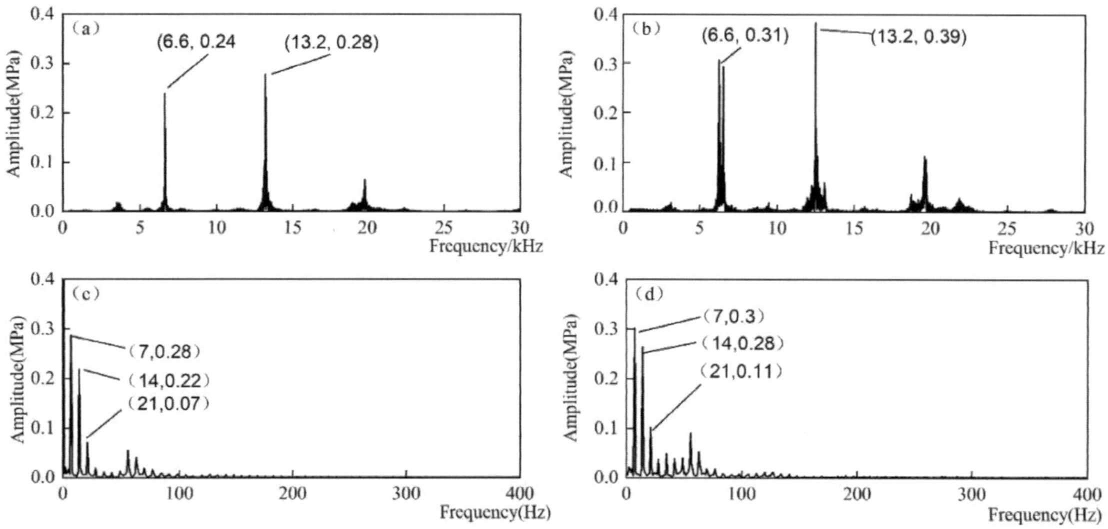

Figure.4 Spectrum of the pressure signal: (a) spectrum at point 2; (b) spectrum at point 4; (c) low frequency spectrum at point 2; (d) spectrum at point 1

The test conditions are the same as above, and the pressure signals of measuring points 1, 2, and 4 are selected for analysis. The measuring point 4 is located on the nozzle inside of the high-pressure tank, to measure the pressure oscillating inside of the nozzle chamber. During the test process, the pressure signals of the measuring points 1, 2, and 4 are acquired synchronously.

Figures 4(a) and (b) show the signal spectrum of points 2 and 4 respectively. It can be seen that the spectra of both signals are composed of high frequency components of about 6.6 kHz and 13.2 kHz. The self-excited frequency is 13.2 kHz, and very close to the acoustic natural frequency of the organ nozzle f^n = 14.9 kHz (Table 1). So the self-resonating water jet achieves the peak resonance at this time. The 6.6 kHz in the spectrum is the sub-harmonic of the self-excited frequency, also known as the half harmonic.

From the spectra, the spectrum structure of measuring point 2 and measuring point 4 is consistent, but the amplitude is slightly different (due to the attenuation of the wave). The measuring point 4 detects the signal inside the nozzle, and the pressure fluctuations inside the nozzle system correlate with target plate pressure fluctuations(Chahine et al. 1983). Figure 4(c) is an enlarged view of the low-frequency part of Figure 4(a), which is compared with the spectrum of measuring point 1 (Figure 4(d)). The fundamental frequencies of them are 7 Hz, which is caused by the fluctuating frequency of high-pressure pump f_p = 6.98 Hz.

The frequency-domain analysis of the pressure signal shows that the spectral analysis of the pressure signal in the pipeline can identify the composition and amplitude of the pressure oscillation characteristics, to characterize the self-resonating water jet.

3.2 Influence of nozzle lip length on frequency characteristic

Tthe influences of the nozzle lip length on frequency characteristic are studied by the above detection method. First, the evolution of self-excited frequency and the corresponding spectra as L_2 increases were given in Figure 5. It is clear from the figure that the self-excited frequency changes significantly with the increase of the L_2/d. In Figure 5(a), at L_2/d = 1, only three peak frequencies are shown and the energy is very concentrated. However, with the increase of L_2/d, the self-excited frequency f^* decrease as shown in Figure.5(b). At the same time, there appear many different frequency components. This resulted from that the self-excited frequency f^* is far from the acoustic natural frequency where no strong resonating occur.

The analysis above shows that the detection method can be used to quantify the performance of self-resonating water jet. Both reliability and accuracy are enough to uncover the influencing mechanism of different factors.

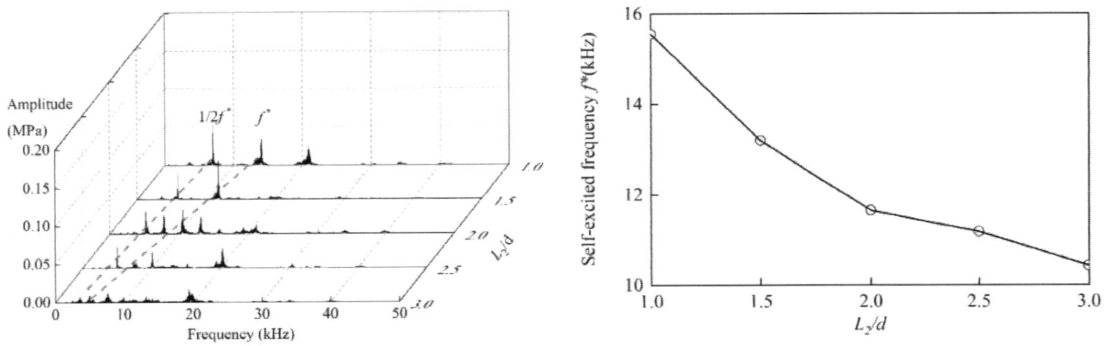

Figure.5 Characteristics changing as a function of L_2/d.(a) Spectrum of the pressure signal spectrum at point 2; (b) self-excited frequency evolution as increasing of L_2/d.

4 CONCLUSIONS

(1) A method for detecting the characteristics of self-resonating water jet based on pressure signal within the pipeline is proposed. Compared with the existing methods, the method moves the pressure sensor from inside to the outside of the high-pressure tank to avoid the influence of confining pressure.

(2) The spectral characteristics of the pressure signal from inside the pipeline are consistent with the signal obtained inside the nozzle cavity, which can characterize the pressure oscillation characteristics of the jet fully.

(3) The detection method was successfully applied for studying the influence of nozzle lip geometry on the characteristic of self-resonating water jet, which verified its reliability and accuracy further.

(4) This method paves the way for the study of self-oscillation jet technology under high confining pressure. It can be used not only as a detection method for the characteristics of self-resonating nozzles in the laboratory but also as a control system for the performance of self-resonating nozzles in the field.

ACKNOWLEDGMENTS

This work is supported by the National Key Technology Research and Development program of China during the 12th Five Year Plan(DY125-14-T-03), the National Natural Science Foundation of China (No. 51774019) and the China Scholarship Council (No. 201806460056).

REFERENCES

Johnson Jr, Virgil E., et al. "The development of structured cavitating jets for deep-hole bits." SPE Annual Technical Conference and Exhibition. *Society of Petroleum Engineers*, 1982.

Li Gensheng et al. "Investigation and application of self-resonating cavitating water jet in petroleum engineering." *Petroleum science and technology* 23.1 (2005): 1-15.

Hu Dong et al. "Analytical and experimental investigations of the pulsed air–water jet." *Journal of fluids and structures* 54 (2015): 88-102.

Li Deng et al. "Effects of area discontinuity at nozzle inlet on the characteristics of high speed self-excited oscillation pulsed waterjets." *Experimental Thermal and Fluid Science* 79 (2016): 254-265.

Ma Fei et al. "Experimental study of self-resonating water jet frequency characteristics." *Journal of Mechanical Engineering* 52.14 (2016): 182-187.

Chahine Georges L., Andrew F. Conn, and Virgil E. Johnson Jr. "*Cleaning and cutting with self-resonating pulsed water jets.*" 2nd US Water Jet Conference. 1983.

Johnson Jr Virgil E., et al. *Research and development of improved cavitating jets for deep-hole drilling*. No. SAND-83-7461. Hydronautics, Inc., Laurel, MD (USA), 1984.

NOMENCLATURE

D, organ-pipe diameter
L, organ-pipe length
d, orifice diameter
L_1, orifice length
L_2, lip length
θ, lip expansion angle
f^*, self-excited frequency
f_n, organ-pipe's acoustic natural frequency
f_p, fluctuation frequency of pump
S_d, Strouhal number
v, jet velocity
M, Mach number
c, sound speed in water
K_n, mode parameter
β, end correction

2019 WJTA Conference and Expo
November 11-13, 2019 • New Orleans, Louisiana
Paper

CHARACTERIZATION OF PHYSICAL FEATURES IMMEDIATELY UPSTREAM OF A NOZZLE INSERT AND THEIR EFFECTS ON WATERJET QUALITY

T. Tormey
StoneAge, Inc.
Durango, Colorado, U.S.A.

ABSTRACT

Many waterblast applications depend on the benefits of a cohesive waterjet. Producing a cohesive waterjet within the size constraints of most applications affords a challenge which relies on many variables. The most critical of these variables affecting waterjet quality are the physical features of the passageways immediately upstream of a nozzle insert. The purpose of this research is to determine, within the size constraints of typical waterblast tooling, the critical parameters of immediate upstream passageways and their contribution to a cohesive waterjet.

Organized and Sponsored by WJTA®

1. INTRODUCTION

Achieving a more cohesive waterjet has been the subject of much study. Without tooling space constraints, the ability to produce a cohesive, far-reaching waterjet is greatly simplified. Few cleaning applications offer such an opportunity. For the applications which are space constrained, understanding the important parameters in optimizing a waterjet are critical and may even determine success or failure.

It's worth noting that not all applications benefit from a cohesive jet. The use of fan jets is a good example of beneficial use of a poor quality, non-cohesive waterjet.

This paper details the important physical features immediately upstream of a nozzle insert to maintain a cohesive jet within the limited space typically available.

2. TEST ARRANGEMENT

Testing was completed by traversing a target across a waterjet at a constant traverse rate, 536 mm/s (21 in/s), with the waterjet normal to the surface of the target throughout the traverse. The target material chosen was blue machinable wax of size 76 mm x 76 mm x 178 mm (3 in x 3 in x 7 in).

Standoff distances from the face of the orifice to the target material surface were 1118 mm (44 in) and 813 mm (32 in). Proper choices of standoff distances to adequately describe waterjet performance are worthy of a paper. However, for this paper, these specific distances were chosen to allow for each test to provide a measurable sample for each test while detailing the difference in performance of varying jet sizes, and facilitate an expedient test program.

To record the results, depth measurements were taken in reference to the flat, undisturbed faces of the machinable wax using a drop gauge and small probe as well as a heaping teaspoon of patience. Ten measurements were taken, each approximately 4 mm apart (.15 in), and arithmetically averaged to determine the depth of cut for each result.

High-pressure water was supplied from a 242 kW (325 hp) triplex pump with pressure measured at the inlet of a 20 mm (0.787 in) supply hose of length 15.24 m (50 ft). All tests were conducted at 68.9 MPa (10,000 psi), filtered to 10 micron (0.0004 in) with carbide nozzles and holders shown in Figure 1.

When flow straighteners were used, they were each 38 mm (1.5 in) long and consisted of the designs shown in Figure 2. Placement of the flow straightener can be seen in Figure 1, with the water passing the flow straightener maintaining the flow straightener's position adjacent to the nozzle insert. Each flow straightener has a simple, geometric division, similar to a pie slice, to detail the effect of additional divisions while limiting uncontrolled variables.

No standard metric for waterjet cohesion exists and so non-space constrained parameters are reported and used for gauging the effects of the physical features upstream of the nozzle insert in space constrained parameters. Non-space constrained parameters include the feeder pipe inside diameters up to 20.7 mm (0.815 in) and feeder pipe lengths up to 914 mm (36 in).

3. FOUNDATIONS FOR TESTED PARAMETERS

Several common practices are used to improve waterjet cohesiveness. This section aims to provide a foundation for the mechanisms behind these improvements to define the important characteristics of flow upstream of a nozzle insert.

3.1 Feeder Pipe Length

Increasing the feeder pipe length has a significant and beneficial effect on waterjet quality. Shavlovsky recommended a minimum ratio of feeder pipe length to feeder pipe inside diameter of 40-50, although he described flow straightener design to be able to reduce the same ratio to 10-15 (A6-87). The goal was "removing circulatory flow arising from the flow twisting and turbulent pulsations" (Shavlovsky, A6-87). One can interpret this as achieving fully developed turbulent flow, with the ratios provided to be the requirements for the flow to stabilize.

Literature regarding the length required for fully developed turbulent flow varies widely. Nikuradse stated that the velocity distributions at length to diameter ratios of 100, 65, and 40 were equal (21). Lien, Monty, Chong and Ooi recommended values of 130 to 150 for fully developed turbulent flow (4). The latter paper details some of the discrepancies on recommended values for fully developed turbulent flow.

Feeder pipe length does not have to be established from an extension arm. A continuous passageway upstream of a nozzle insert can be formed from a tap drill and result in an improvement to waterjet quality. These improvements are typically small since the length to diameter ratio lies well below any of the previously stated values.

Regardless of the incongruities, the entry length, or the length required to establish fully developed turbulent flow is related to the hydraulic diameter and the flowrate and any of the published minimum lengths are usually significantly longer than most waterblast applications can accommodate.

3.2 Feeder Pipe Diameter

Though the Reynolds number lies well within the turbulent region within the feeder pipes of waterblast systems, reductions in Reynolds Number are correlated with the benefits of increasing feeder pipe diameter.

Reynolds Number:

$$\mathrm{Re} = \frac{\mathrm{uL}}{\nu} = \frac{\mathrm{QD_h}}{\mathrm{A}\nu} \qquad \text{(Equation 1)}$$

With an increase in feeder pipe diameter, the Reynolds Number is reduced because the velocity of the fluid within the feeder pipe is reduced. Shavlovsky noted this also and recommended maximizing the ratio of nozzle diameter to feeder pipe inside diameter, with values up to 9-10 suggested (A6-88). Wright found similar benefits of this ratio with an optimal value of 5-13, depending on the length of the feeder pipe (4).

Another important relationship is between turbulence and power. With increasing nozzle size comes increasing flow and, since flow is a linear variable in both the Nozzle Power equation and the Reynolds Number equation, the ratio of Nozzle Power to Reynolds Number is constant for each feeder pipe diameter. In other words, the power applied to the system and the degree of turbulence in the system change at the same rate. Some testing has shown peak effective nozzle sizes for certain flow regimes that would indicate the Nozzle Power to Reynolds Number ratio could be used as an Efficiency Index. As the Efficiency Index increases, the peak effective nozzle diameter is increased, and more power may be applied for the same amount of turbulence in the system.

Nozzle Power:

$$P = \frac{Qp}{C} \qquad \text{(Equation 2)}$$

Efficiency Index:

$$E.I. = \frac{P}{Re} = \frac{pAv}{D_h C} \qquad \text{(Equation 3)}$$

As with feeder pipe length, the diameter of feeder pipes is typically limited by the application and the pressure which, again, typically dictate maximum sizes for tooling.

3.3 Flow Straighteners

Using flow straighteners has two beneficial effects:

1. Water flow is more fully developed at the nozzle by decreasing the hydraulic diameter. Typically done by increasing feeder pipe length.
2. Turbulence is reduced by maintaining fluid velocity while decreasing the hydraulic diameter. Typically done by increasing feeder pipe diameter.

Flow straighteners "reduce the diameter of the effective tube by dividing the channel into a series of smaller segments" (Summers, 41). Essentially, the hydraulic diameter the water passes through decreases without significantly raising the velocity of the water within the feeder pipe.

A further reduction of the Reynolds Number can be achieved by optimizing the number of flow straightener divisions. The acting area of the feeder pipe described using the number of passages and wall thickness of the flow straightener is seen in Equation 4.

Reduced Cross-Sectional Area:

$$A_r = \frac{\pi}{4}D_h^2 - \frac{D_h}{2}Nw \approx \frac{\pi}{4}ND_f^2 \qquad \text{(Equation 4)}$$

Some liberty with the equation is taken here since dividing the circle of the feeder pipe inside diameter does not result in circular regions.

Substituting the Reduced Area into Equation 1 and some simplification gives:

$$Re = \frac{Q}{\frac{\pi}{4}vND_f} \qquad \text{(Equation 5)}$$

Equation 5 shows, for each feeder pipe diameter, there exists a minimum Reynolds Number formed by the inverse relationship between number of flow straightener divisions (N) and the size of the passageways formed by the divisions (D_f).

The diameter of the passageways formed is a function of the number of divisions. Further reduction in the equation allows for the following:

$$Re = \frac{Q}{\frac{\pi}{4}v\sqrt{ND_h^2 - \frac{2}{\pi}N^2D_hw}} \qquad \text{(Equation 6)}$$

Substitution of the hydraulic diameter allows the partial derivative of the Reynolds Number with respect to flow straightener divisions ($\partial Re/\partial N$) to be taken to produce the optimal number of passageways. The author won't spoil all the fun for the reader, though correct answers of this partial derivative may be submitted to the author in exchange for a free haircut by the author.

It should also be noted that the flow straightener walls have a displacement thickness which should be considered. The displacement thickness forces a reduction in the effective cross-sectional area, acting as a thicker wall than the physical flow straightener actually has. Much literature exists for displacement thickness of flat plates and moderately high Reynolds numbers, but the parameters typically encountered in waterblast tooling don't allow these to be used easily.

Because of their cost, size, and effectiveness, flow straighteners are ubiquitous in the waterblast industry. Typically found in the nozzle inserts themselves, the ability of a flow straightener to improve waterjet quality in a very small space makes them a unique and powerful component.

3.4 Individual Passageways

Many tools use nozzle inserts which are fed by individual passageways to improve jet quality. Similar to increasing feeder pipe length, individual passageways allow for flow to be more developed just upstream of the nozzle insert.

Packing enough individual passageways into an appropriate size for waterblast tooling can limit this technique for improving waterjet quality. Maintaining a reasonably-sized tool results in a reduction in passageway ID and a subsequent increase in turbulence upstream of the nozzle insert. Also, the overall length of the passageway may not be contributing to fully developed flow since the curves of the passageway induce turbulence within.

High-quality waterjets may be created with this method, but it requires a difficult balance of many parameters.

3.5 Surface Roughness

Having a rough surface upstream of a nozzle insert can, just like a golf ball, produce a thin boundary layer that is beneficial to waterjet quality. However, producing a feeder pipe or other tooling with a prescribed surface finish likely isn't an economically sound pursuit and isn't common in the waterblast industry.

4. TESTED PARAMETERS

4.1 Feeder Pipe Length

Results in Figure 3 and Figure 5 show that the performance gains from feeder pipe length behave logarithmically and the largest performance differences are in the shorter lengths. Each feeder pipe diameter saw the majority of performance gains, 80%, in feeder pipes shorter than 304 mm (12 in). Something important to note is that the results don't indicate any significant decrease in performance with a feeder pipe longer than the optimal length.

Where possible, feeder pipe length should be increased to reduce standoff distance and improve waterjet quality.

4.2 Feeder Pipe Diameter

Results in Figure 7 show that the performance gains from a large feeder pipe diameter are significant, especially when coupled with a long feeder pipe.

More importantly, shown in Figures 3 and 5 are the ability to benefit from more power being applied to the system because of larger Efficiency Index values. The 11.1 mm (.438 in) feeder pipe did not see better performance with the 3.2 mm (.125 in) orifice, whereas the 20.7 mm (.815 in) feeder pipe has higher Efficiency Index values and did perform better with the larger orifice.

Particularly in high-flow applications, feeder pipe diameter should be increased to improve waterjet quality.

4.3 Flow Straighteners

Flow straighteners with simple geometric divisions are shown in Figure 2 and were shown to have a significant benefit to waterjet quality as seen in Figure 3, raising the performance of the feeder pipe by 37%. The effects of increasing the number of flow straightener divisions is shown in Figures 8 and 9 indicating that, depending on orifice size, the optimal number of divisions may be greater than six for a feeder pipe of diameter 11.1 mm (.438 in).

Similar to the elevated values of the Efficiency Index from a large feeder pipe diameter correlating to better performance with larger nozzles, more flow straightener divisions raised the Efficiency Index and the trend was improved performance with larger nozzles. This point is especially important for high-flow applications; by adding the appropriate flow straightener, a cohesive jet can be made with a small diameter feeder tube and a large diameter orifice.

4.4 Individual Passageways

Limited testing was completed on components with an individual passageway feeding a nozzle insert. Benefits of individual passageways were seen in some, but not all, cases. The individual passage is not sufficient to improve waterjet quality.

The magnitude of benefits to waterjet quality were comparable to a feeder pipe with a similar straight length directly upstream of the nozzle insert. Also, smooth transition features between the high-pressure supply and the start of the individual passageways had no measurable effect on waterjet quality.

4.6 Surface Roughness

Limited testing was completed on the surface roughness of parts immediately upstream of the nozzle insert. The results showed a slight benefit when there was some degree of roughness to the wall of the feeder pipe.

5. CONCLUSIONS

Many features immediately upstream of a nozzle insert influence waterjet quality, though there are three that have a significant effect:

1) Feeder pipe length – through aiding in flow development within the pipe.
2) Feeder pipe inside diameter – through reduction in turbulence within the pipe.
3) Flow straighteners – through reduced distance required for developed flow and reductions in turbulence within the feeder.

While it is not always practical to use large feeder pipes, there is no significant penalty for making the feeder pipe as long as the application allows. Also, feeder pipe length should be considered first, as the benefits to additional length are greater than the benefits from a slightly larger diameter.

For space constrained applications, flow straighteners provide the most effective method to improve waterjet quality, especially when high flow causes extreme turbulence upstream of the orifice. Performance can be recovered in a small space and large orifices can be used more effectively.

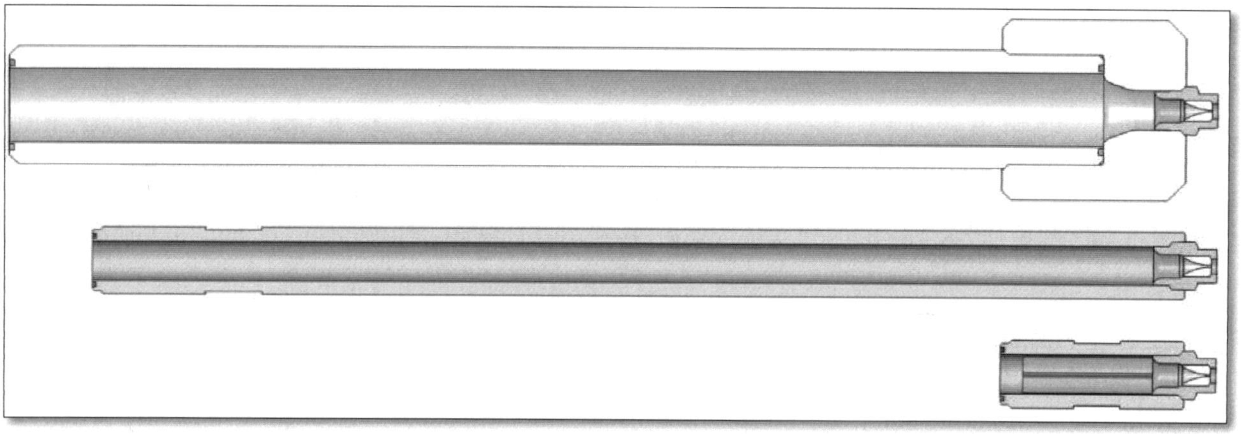

Figure 1. Top: Ø20.7 mm by 305 mm (Ø.815 in x 12 in) feeder pipe with carbide nozzle holder. Middle: Ø11.1 mm by 305 mm (Ø.438 in x 12 in) feeder pipe with carbide nozzle holder. Bottom: Ø11.1 mm by 51 mm (Ø.438 in x 2 in) feeder pipe with carbide nozzle holder and flow straightener installed.

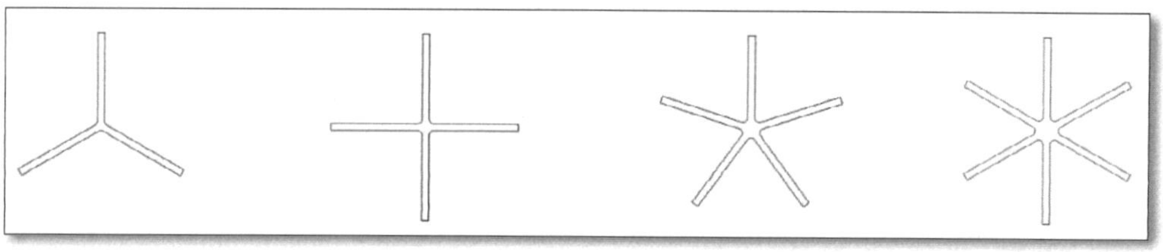

Figure 2. Flow straightener cross sections showing styles with 3, 4, 5, and 6 divisions.

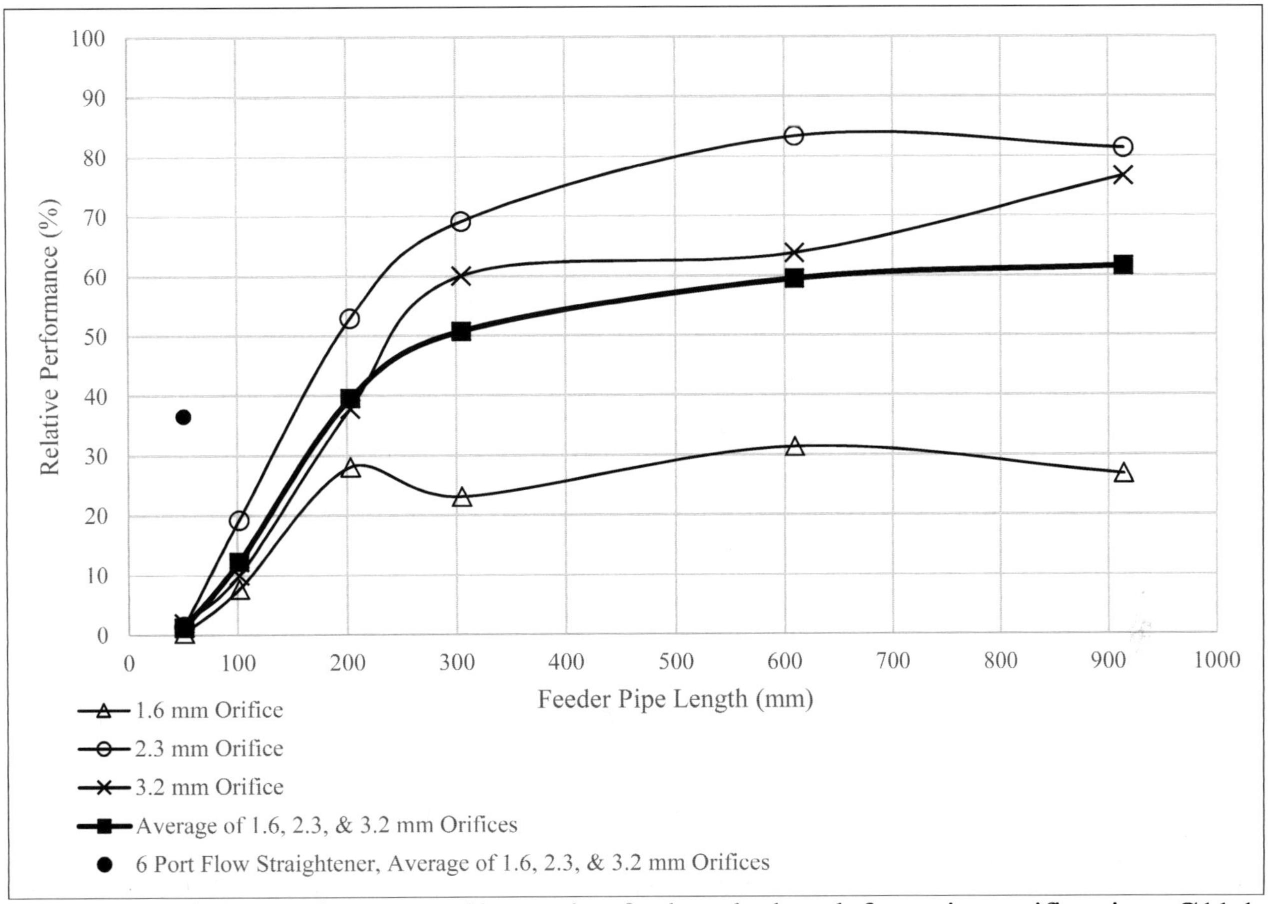

Figure 3. Relative performance of increasing feeder tube length for various orifice sizes, Ø11.1 mm ID feeder pipe. Performance relative to 3.2 mm (.125 in) orifice with Ø20.7 mm by 914 mm (Ø.815 in x 36 in) feeder pipe. Note stabilization of performance with increasing length as well as increased performance with a flow straightener.

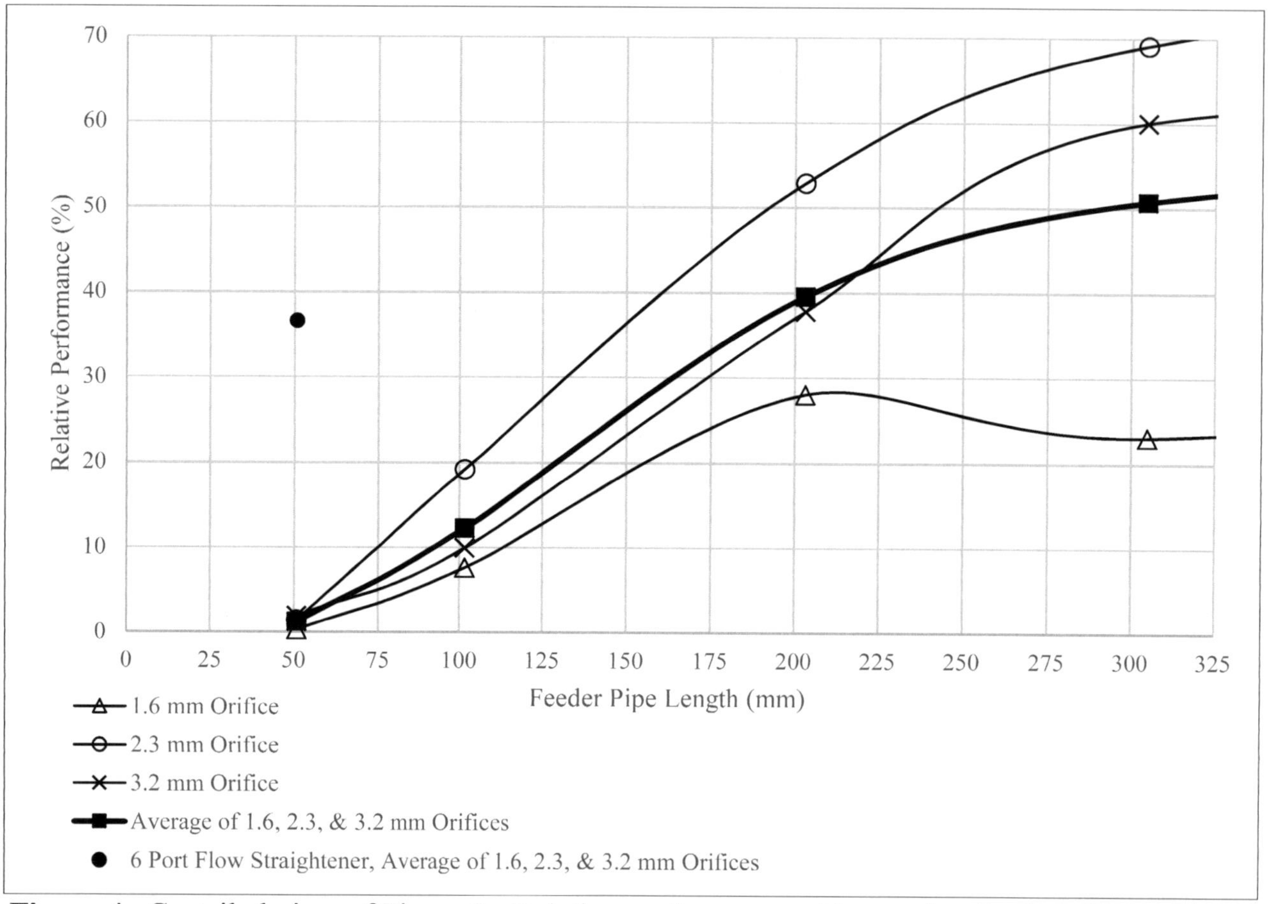

Figure 4. Curtailed view of Figure 3. Relative performance of increasing feeder tube length for various orifice sizes, Ø11.1 mm ID feeder pipe. Performance relative to 3.2 mm (.125 in) orifice with Ø20.7 mm by 914 mm (Ø.815 in x 36 in) feeder pipe. Note stabilization of performance with increasing length as well as increased performance with a flow straightener.

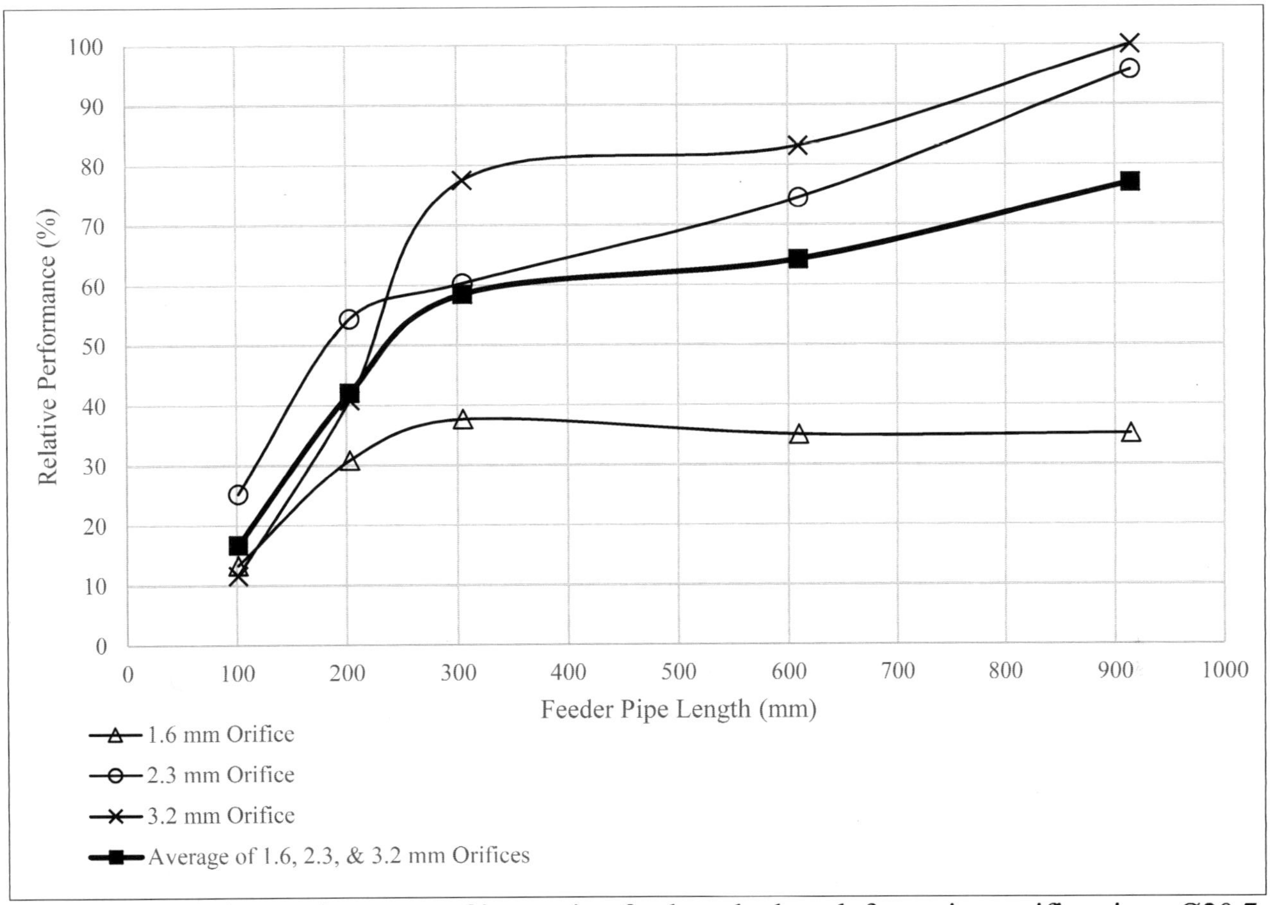

Figure 5. Relative performance of increasing feeder tube length for various orifice sizes, Ø20.7 mm ID feeder pipe. Performance relative to 3.2 mm (.125 in) orifice with Ø20.7 mm by 914 mm (Ø.815 in x 36 in) feeder pipe. Note stabilization of performance with increasing length.

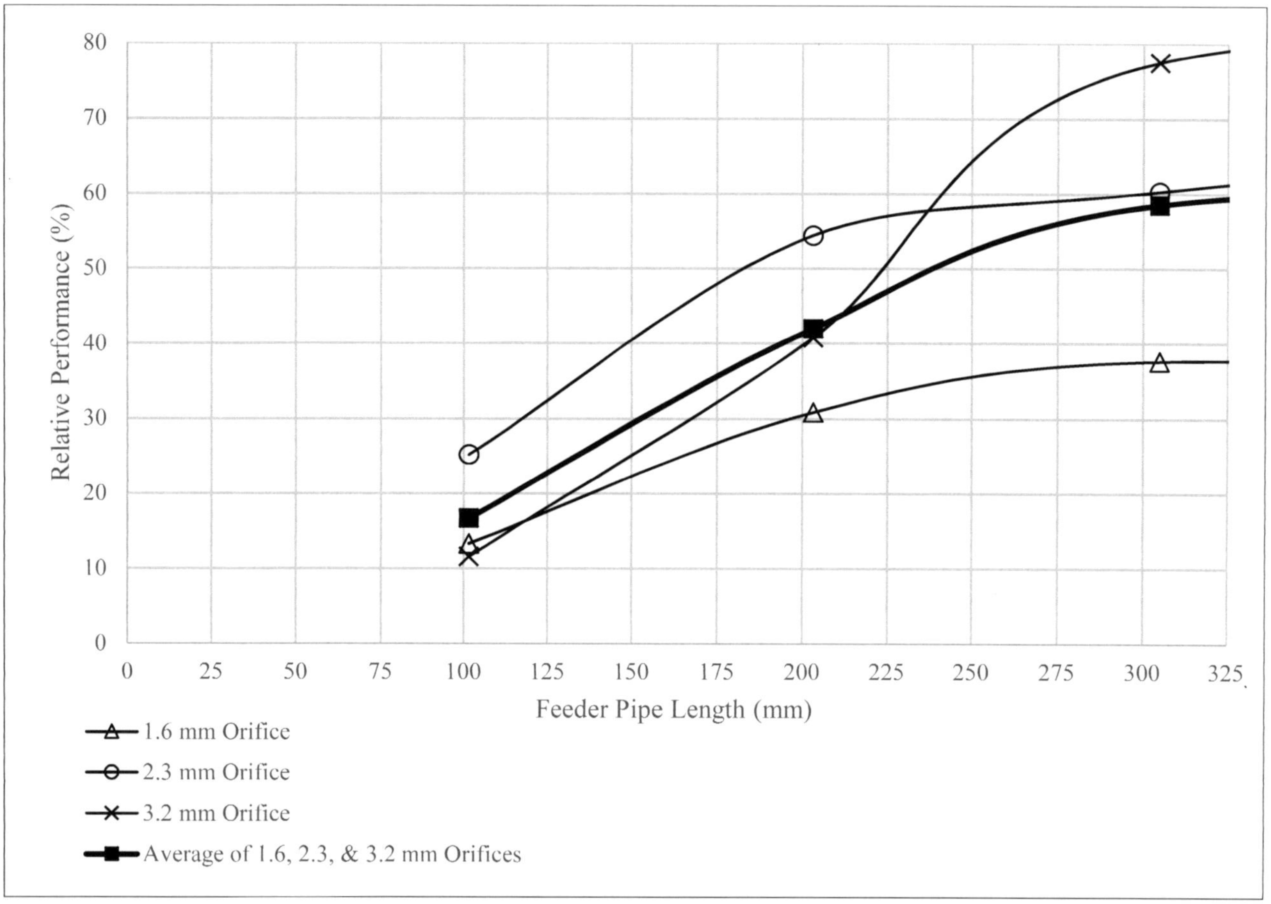

Figure 6. Curtailed view of Figure 5. Relative performance of increasing feeder tube length for various orifice sizes, Ø20.7 mm ID feeder pipe. Note stabilization of performance with increasing length.

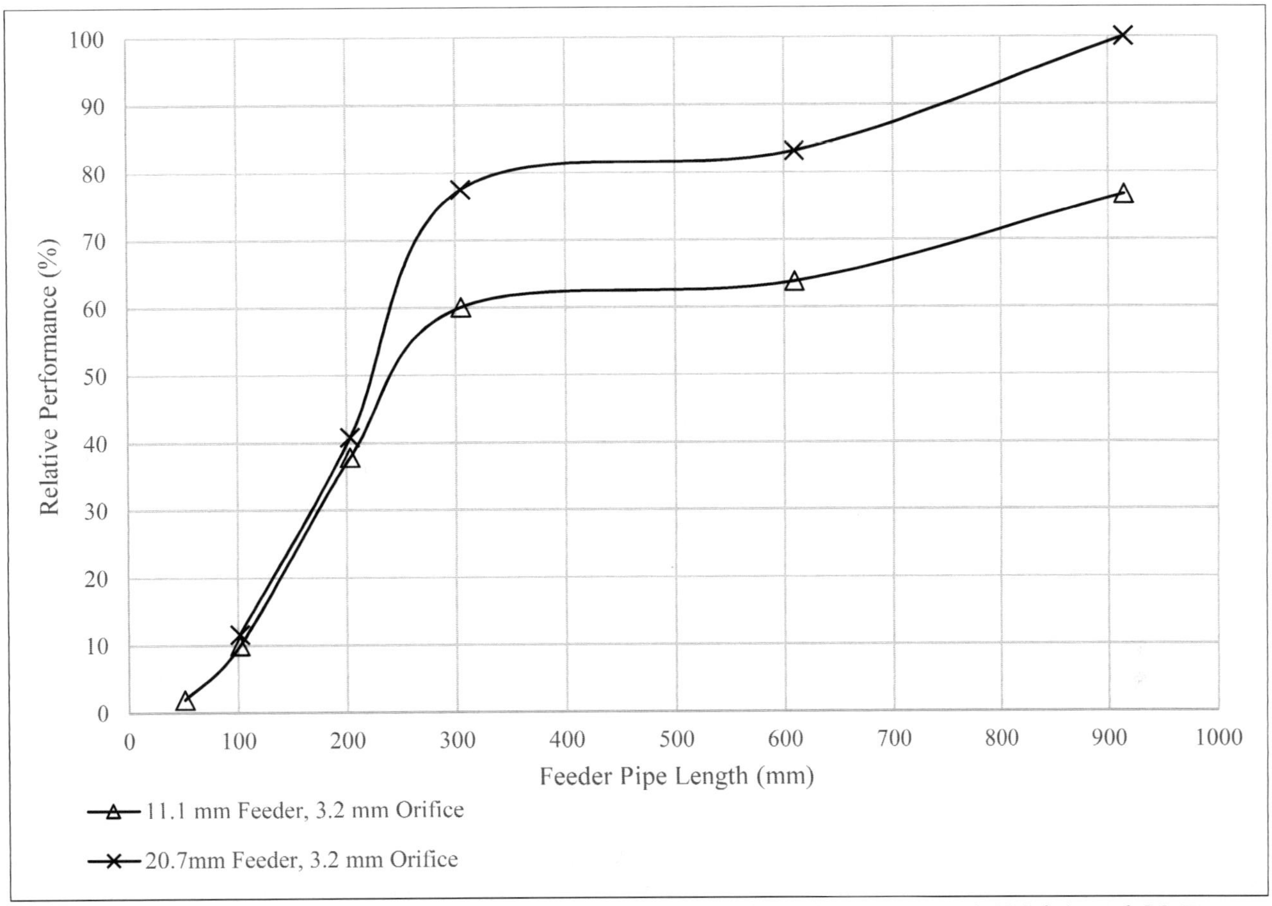

Figure 7. Relative performance of 3.2 mm (.125 in) orifices in 11.1 mm (.438 in) and 20.7 mm (.815 in) feeder pipes of varying length. Performance relative to 3.2 mm (.125 in) orifice with Ø20.7 mm by 914 mm (Ø.815 in x 36 in) feeder pipe. Note similar performance with short feeder pipes and improved performance with a larger diameter feeder pipe as the length increases.

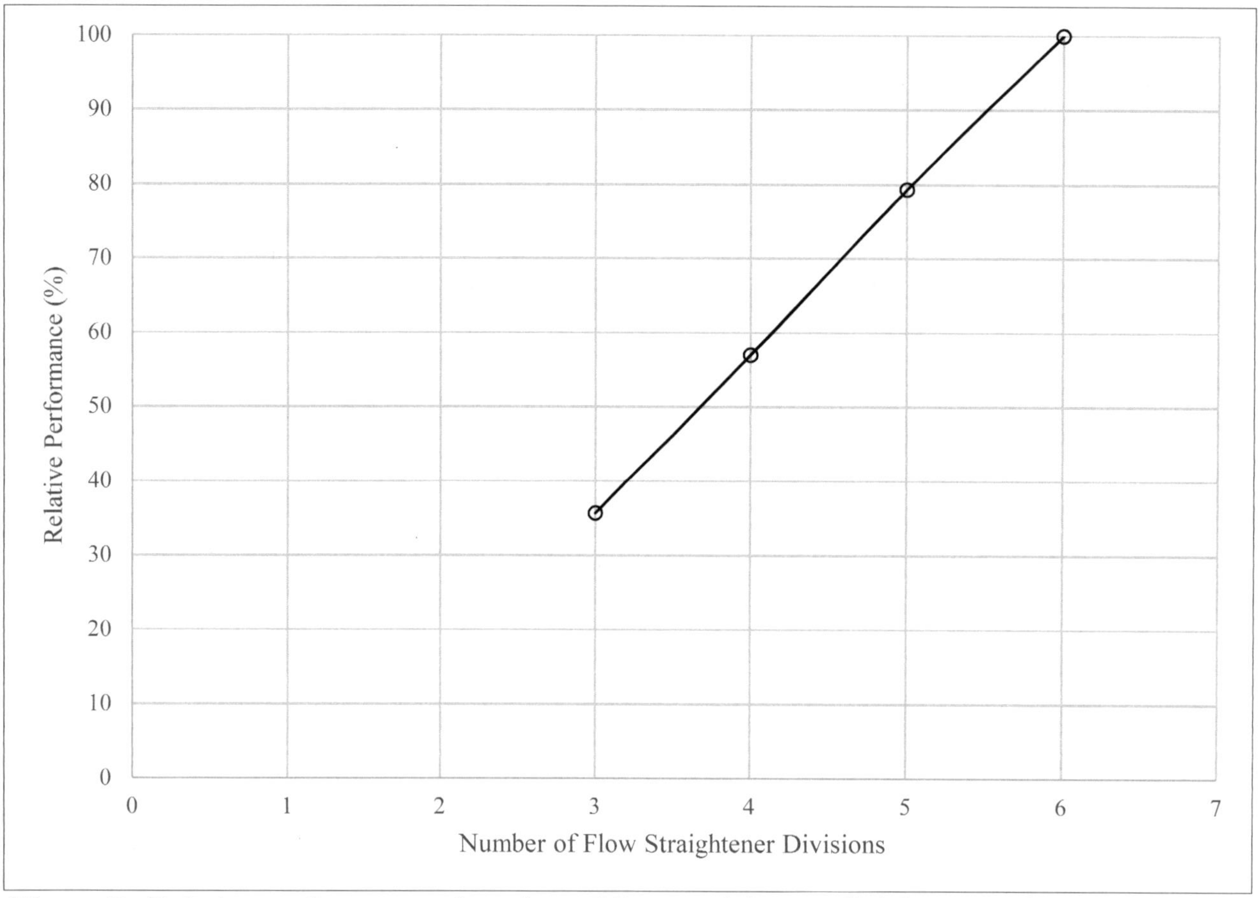

Figure 8. Relative performance of number of flow straightener divisions. Performance relative to average of 1.6, 2.3, and 3.2 mm (.063, .090, .125 in) orifices with Ø11.1 mm by 51 mm (Ø.438 in x 2 in) feeder pipe. Note the trend of increasing performance with increasing divisions as well as the slope of the line suggesting that the trend may continue past six divisions.

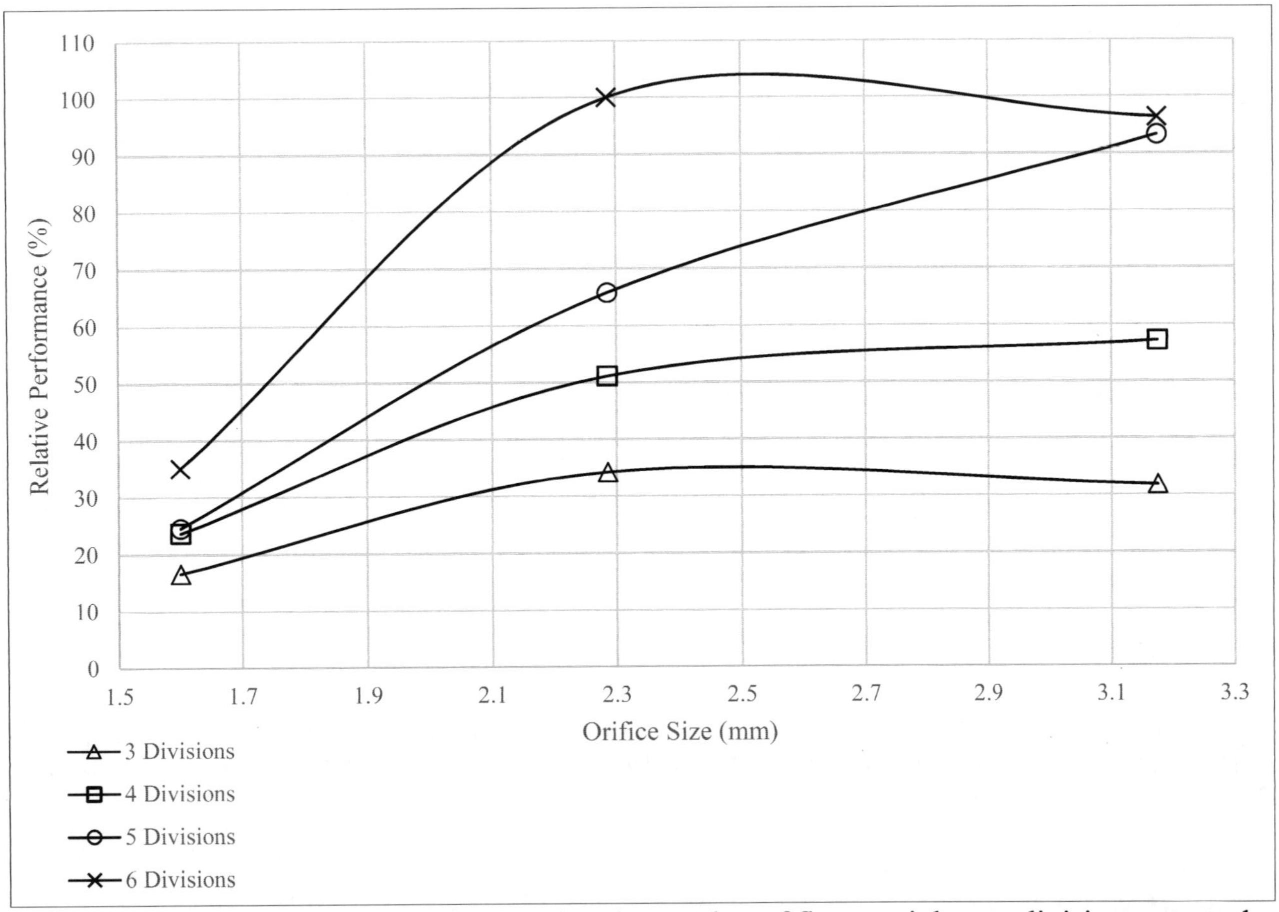

Figure 9. Relative performance of increasing the number of flow straightener divisions vs nozzle size. Performance relative to a 2.3 mm (.090 in) orifice with a six-division flow straightener installed in a Ø11.1 mm by 51 mm (Ø.438 in x 2 in) feeder pipe. Note that the optimal orifice size may be larger than 3.2 mm (.125 in) for the flow straighteners with four and five divisions.

6. REFERENCES

Lien, K, Monty, J.P., Chong, M.S. and Ooi, A. *The Entrance Length for Fully Developed Turbulent Channel Flow*, 15th Australasian Fluid Mechanics Conference, 2004, Sydney, Australia, Curran Associates, 2013.

Nikuradse, J. *Laws of Turbulent Flow in Smooth Pipes*, Research in the Field of Engineering, 3, 1932, 1-36, Translated to NASA TT F-10, 359, 1966.

Pritchard, Philip, and Leylegian, John. *Introduction to Fluid Mechanics*. 8th ed., John Wiley & Sons, 2011.

Shavlovsky, D.S., "Hydrodynamics of High Pressure Fine Continuous Jets." *First International Symposium on Jet Cutting Technology*, pp A6-81-A6-92, BHRA Fluid Engineering, Cranfield, UK, 1972.

Summers, David. "Flow Straighteners." *Cleaner Times*, July 2005, pp. 40-42

Wright, Doug. "The Effect of Upstream Feeder Tube Diameter and Length on Waterjet Performance in Cleaning Applications." *WJTA American Waterjet Conference*, November 2-4, 2015, New Orleans, Louisiana, 2015

7. NOMENCLATURE

Re = Reynolds number
u = velocity of fluid
L = characteristic length
ν = kinematic viscosity
Q = volumetric flowrate
A = cross-sectional area of flow pathway
A_r = reduced cross-sectional area of flow pathway. Reduced by wall thickness of flow straightener
P = power
p = pressure
C = constant
E.I. = efficiency index
D_h = hydraulic diameter of feeder pipe
D_f = hydraulic diameter formed by flow straightener
N = number of flow straightener divisions
w = wall thickness of flow straightener

AUTHOR INDEX

WaterJet Technology Association-Industrial & Municipal Cleaning Association
906 Olive Street – Suite 1200
St. Louis, MO 63101-1448
USA

ISBN 978-1-7138-2818-1

9 781713 828181

CURRAN ASSOCIATES INC.
proceedings.com